Small Wild Cats: *The Animal Answer Guide*

Small Wild Cats

The Animal Answer Guide

James G. Sanderson and Patrick Watson

The Johns Hopkins University Press Baltimore

Printed in the United States of America on acid-free paper
9 8 7 6 5 4 3 2 1

The Johns Hopkins University Press
2715 North Charles Street
Baltimore, Maryland 21218-4363
www.press.jhu.edu

Library of Congress Cataloging-in-Publication Data

Sanderson, James G., 1949–
Small wild cats : the animal answer guide / James G. Sanderson and Patrick Watson.
p. cm.
Includes bibliographical references and index.
ISBN-13: 978-0-8018-9884-6 (hardcover : alk. paper)
ISBN-13: 978-0-8018-9885-3 (pbk. : alk. paper)
ISBN-10: 0-8018-9884-6 (hardcover : alk. paper)
ISBN-10: 0-8018-9885-4 (pbk. : alk. paper)
1. Felidae—Miscellanea. I. Watson, Patrick, 1947– II. Title.
QL737.C23S235 2011
599.75—dc22 2011000460

A catalog record for this book is available from the British Library.

Special discounts are available for bulk purchases of this book. For more information, please contact Special Sales at 410-516-6936 or specialsales@press.jhu.edu.

The Johns Hopkins University Press uses environmentally friendly book materials, including recycled text paper that is composed of at least 30 percent post-consumer waste, whenever possible.

Spontaneous activity and sensitivity are the special characteristics of animal life, and the cats are well endowed with both these powers. The perfection of their organs of movement and that of the very substance of their bones and muscles, as well as the great perfection of their special senses is widely recognized. Some have objected, however, that the activities and sense perceptions of certain other beasts are, in their own various ways, as highly developed as are those of the Felidæ. Only through the possession of perfectly formed bones and muscles, of a delicate sense of hearing, or of far-reaching vision do antelopes, hares, and such creatures escape their carnivorous pursuers. But then, they use their organization for escape.

The organization of the cats may then be deemed superior, not only because it is excellent in itself, but because it is fitted to dominate the excellences of other beasts. Thus considered, the Carnivora rank first amongst mammals, and the cats rank first amongst the Carnivora. Moreover, much may be said in favor of cats being the highest of mammals if man is considered merely in his animal capacity—in which alone he can be brought into comparison with other organisms. But whether or not this eminence be allowed to the cat, there can be no question that it is the most highly developed form of carnivorous mammalian life—the most perfect embodiment, the essence, of a beast of prey. Such, then, is certainly the cat's place in nature.

St. George Mivart, *The Cat*, 1892

Contents

Acknowledgments xi
Introduction xiii

1 Introducing Small Cats 1
What are small cats? 1
What is the difference between small cats and big cats? 7
How many kinds of small cats are there? 9
Why are small cats important? 10
Where do small cats live? 10
Why are there only domestic small cats on Madagascar, Australia, New Guinea, and New Zealand? 12
How are small cats classified? 13
What characterizes the major groups of small cats? 16
When did small cats evolve? 18
What is the oldest small cat fossil? 19

2 Form and Function of Small Cats 20
What are the largest and smallest small cats? 20
How fast does a small cat's heart beat? 22
Can small cats see color? 24
Are small cats capable of retracting their claws? 27
Can small cats swim? 27
How far can small cats jump? 28
Are small cats capable of climbing? 31
Are small cats' tails the same? 32
Do small cats have whiskers? 32
How sensitive is a small cat's sense of smell? 36
What is special about a cat's sense of taste? 37
Can you determine whether a fossil small cat was terrestrial or arboreal? 39

3 Small Cat Colors 40
Can small cats be black? 40
What causes the different coat colors of small cats? 40
Why does the length and thickness of small cat's coat vary? 41
How are hair colors determined genetically? 43

Is there a reason for the patterns on the coat? 44
Are there age-related differences in coat color? 46
Are there seasonal changes in coat color? 47
Is there much geographic variation in small cat species? 47

4 Small Cat Behavior 50
Are small cats social? 50
Do small cats fight? 51
How smart are small cats? 51
Do small cats play? 53
Do small cats talk? 54
How do small cats avoid predators? 54

5 Small Cat Ecology 55
Where do small cats sleep? 55
Do small cats migrate? 56
How many small cat species can coexist? 58
Are small cats equally distributed throughout the world? 59
How do small cats survive in the desert? 60
How do small cats survive the winter? 61
Do small cats hibernate? 62
Do small cats have enemies? 63
Do small cats commit infanticide? 63
Do small cats get sick? 64
How do small cats influence vegetation? 64

6 Reproduction and Development of Small Cats 66
How do small cats reproduce? 66
How long are female small cats pregnant? 67
Where do mother small cats give birth? 68
How many babies do small cats have? 71
Are all littermates equally related? 71
How long do female small cats nurse their young? 72
How fast do small cats grow? 72
How long do small cats live? 73

7 Foods and Feeding of Small Cats 75
What do small cats eat? 75
How do small cats hunt? 78
Do small cats hide or bury food? 81
How often do small cats eat and drink? 83
Do small cats scavenge? 83

8 Small Cats and Humans 84
Do small cats make good pets? 84
Should people feed small cats? 85
Do small cats feel pain? 85
What do I do if a find an injured or orphaned small cat? 86
How can I become a better observer of small cats? 87
How do I know whether I have small cats in my backyard? 88
Why are small cats important? 89

9 Small Cat Problems (from a human viewpoint) 90
Are small cats pests? 90
How are small cats kept away from people, livestock, and poultry? 91
Are small cats vectors of human disease? 92

10 Human Problems (from a small cat's viewpoint) 94
Are small cats endangered? 94
Will small cats be affected by global warming? 97
Are small cats ever invasive species? 98
Do people hunt and eat small cats? 99
Why are some small cat skins so valuable? 101
Why do small cats get hit by cars? 102

11 Small Cats in Stories and Literature 103
What roles do small cats play in religion and mythology? 103
Are some small cats considered bad luck? 108
What roles do small cats play in popular culture? 109
Are there popular sayings about small cats? 112
How are small cats incorporated into poetry? 113
How are small cats incorporated into literature? 118

12 "Small Catology" 121
Which species are best known? 121
Which species are least known? 123
How do scientists recognize individual small cats? 126

Appendix: Small Cats of the World 129
Bibliography 131
Index 139

Acknowledgments

Many friends and colleagues share our enthusiasm for small wild cats. Our interest has been brought about by an increase in our understanding of small cats worldwide. As our knowledge expands, concern regarding the future of small cats is justified. The number of people interested in the conservation of small cats has likewise grown and will continue to grow as more people become familiar with these small, beautiful, and secretive top predators. We hope our book brings the world of small wild cats to an even greater audience and leads to an increase in awareness of the plight faced by these cats in their everyday lives. Understanding leads inevitably to appreciation and ultimately to protection.

The information presented here represents the accumulated knowledge gained by many past and present naturalists and scientists from all over the world. It is impossible and unnecessary to thank them all. However, some deserve special thanks for influencing our understanding of the world we live in and enabling us to work on small cat conservation more effectively. JGS thanks Ian Anderson, Kennon Dickson-Hudson, the late John Eisenberg, Grant Harris, Larry D. Harris, Darla Hillard, Rodney Jackson, Charlie Knowles, Mike Moulton, Patsy and Cleve Moler, Stuart Pimm, Fiona and Mel Sunquist, Ron Thompson, and all his colleagues and friends of the Feline Conservation Federation, IUCN Cat Specialist Group, Small Wild Cat Conservation Foundation, University of Arizona Wildcat Research and Conservation Center, and Wildlife Conservation Network.

During our efforts to write about small cats, we were guided by Vincent J. Burke, executive editor at the Johns Hopkins University Press, who succeeded in helping us overcome whatever hurdles were in our path.

PW thanks two special friends, Michael Hutchins and the late Kailash Sankhala, for their support, knowledge, and help.

Finally, JGS thanks his spouse and partner in conservation, Joan L. Morrison, for sharing his enthusiasm for the natural world where both the mighty and the small coexist in harmony and want nothing more than to be left alone to live out their lives.

Introduction

Whether pondering the family pet or reading of one in its natural state, people have always had a fascination with cats. Large or small, all cats are members of the Great Family of Cats, the Felidae. Of the 37 known species of cats that inhabit the planet, most are referred to as the "small cats." These small cats include the familiar domestic variety, which scientists usually refer to as *Felis catus*. We are all familiar with the big cats such as the lion, tiger, cheetah, leopard, and puma or mountain lion; however, very few could name many of the other 30 species, except perhaps bobcats and lynx. But the diversity of small wild cat species is quite remarkable and includes fascinating animals such as the black-footed cat, the flat-headed cat, the Andean cat, and the bay cat.

Why a guide to small cats? The easy answer is that they are beautiful and mysterious and add a diversity of behaviors and habits not found in the big cats. And they intrigue many of us. Unlike the lion and some domestic and feral cats, all of the small cat species live a solitary, secretive life. Many are rare and in danger of becoming extinct, and some are so rare and difficult to study that we can describe only their appearance and approximate geographic range. Some have never been studied in the wild.

Unlike the big cats, small cats have been little studied in the wild. Over time, myths and legends about many of the species have grown. However, we now know enough to provide answers to many questions both the scientist and the naturalist have been asking. One fact has emerged as our knowledge base has broadened: small wild cats are nature's most perfect predators. This guide has been created to answer the many questions that arise about these small but unique creatures.

Apart from containing the known facts on small cats, this book draws on the experience and knowledge of people who live and work in the field with the animals. In these pages you will learn about the water-loving fishing cat of Southeast Asia, which has claws that don't fully retract, and the golden cat of Asia, with its powerful limbs and muscular body. These and other features of the small cats allow us to ask and answer many of the "why" questions that make biology fun. We have tried to select questions that we believe you have, such as, Do small wild cats socialize and live in groups? How rare are they? What do they eat? How do they hunt? And what can I do help them?

One of the important aspects about this book is its ability to convey in a clear and precise way the many differences between species that at first

When it comes to taking life easy, small cats like their larger cousins know exactly how to relax. Here we see an ocelot resting in a typical cat-like pose. Photo by Tom Smylie, USFWS

Combining all the feline qualities of its wild cousins, the domestic cat is the most numerous and widespread member of the cat family. In a relaxed mood, this domestic cats pose mirrors that of the ocelot. Photo © Patrick Watson

sight may seem similar. In many ways these cats appear to be simply smaller versions of their larger cousins. The reader will soon learn, however, that the small cats are quite different from the large species and that often there are features unique to some or even one species of small cat. We show that small cats have a compelling story to tell. We describe their origins and evolution, classification and lineage, and biology and behavior. Answers are provided about their form and function, reproduction and development, contact with humans, and the likelihood that they will make it through the twenty-first century.

Although the authors have endeavored to answer most of the questions posed on the subject, they do so in the hope that the guide provokes further interest and discussion. But more important, we trust the publication of this book will raise awareness and stimulate action regarding the plight of these increasingly rare and endangered creatures.

Small Wild Cats: *The Animal Answer Guide*

Chapter 1

Introducing Small Cats

What are small cats?

Small cats are one of the most widely known and recognized mammals on earth, inhabiting every continent including Antarctica. One member of the group, the domestic cat, even makes its home in our own home and is a familiar member of the family. Cats are so familiar to humans that with even a brief glance everyone recognizes them. Often, however, only an expert can precisely identify the species of a particular cat. Small cats, like their larger cousins, are members of the Felidae family, whose common ancestor arose around 35 million years ago and gave rise to the 37 species of cats recognized in this family today.

All cats share certain common morphological (anatomical) characters, such as a reduced number of teeth, which can be used to identify fossils, and phenotypic (molecular) characters that scientists use to organize and group living species. Although there are important differences in both morphological and phenotypic characters among the world's cats, cats are recognized to have evolved conservatively. Modifications and improvements throughout their evolutionary history have been small and often subtle. Apparently, improving on nature's most perfect terrestrial predators does not require major changes in morphology. However, there are both morphological and phenotypic similarities and differences from one small cat to the next. Some of these similarities and differences are discussed below.

TEETH AND SKULL. Like all wild cats, small cats are *hypercarnivores*—obligate meat eaters. No other predators are so tied to a predatory way of life. This lifestyle has a profound influence on the teeth and skull. Small

Table 1.1. Small cats scientific classification

Domain	Eukaryota
Kingdom	Animalia
Phylum	Chordata
Class	Mammalia
Order	Carnivora
Family	Felidae
Subfamily	Felinae

cats, like their larger relatives, have a reduced number of teeth compared to other carnivores. All small cats have just 28 or 30 teeth, far fewer than other carnivores such as bears and wild dogs, which have a more omnivorous diet and 42 teeth. The snout of a small cat is foreshortened because the jaw holding these teeth does not need to be as large as a bear's or wild dog's, which extends from the face. The teeth of all carnivores are used for securing and consuming food. Evolution has produced finely tuned instruments precisely adapted to these important tasks.

The upper jaw teeth differ from the lower jaw teeth, though the teeth are similar from side to side. The front teeth on the lower and upper jaw, the incisors, are both simple and small. The incisors stand side by side, and viewed from the front are six across on the lower and on the upper jaw, just as in humans. The incisors are used for holding prey and stripping meat from bone.

The canines on the lower jaw stand at either end of the incisors. On the upper jaw, the canines are slightly separated from the incisors by a small space or *diastema*. When the jaw closes, the lower canines fit tightly in the diastema. The exposed part of a canine tooth is called the *crown*, and the part of the canine tooth embedded in the jawbone is the *fang*. The crown is curved and sharply pointed. The front-facing surface has a prominent vertical groove, and the rear of the canine has a similar though more shallow vertical groove. These four large, prominent canine teeth are used to swiftly penetrate the skin or hide of a prey animal so that a fatal bite, sometimes to the back of the neck, can be swiftly delivered. The canines separate the neck vertebrae of the prey animal and sever the spinal cord. The six lower and six upper incisors and two lower and two upper canines account for 16 teeth, more than half the total number.

The incisor and canine teeth are in the front of the cat's mouth. Behind these and further back in the jaw are other specialized teeth. The next tooth in the upper jaw, missing in the lower jaw, is a very small tooth separated by an *interspace* behind the canine. This tooth is the first *premolar*; it has a small crown and a single fang that secures its place in the jawbone. The second premolar and the remaining teeth appear in both the upper

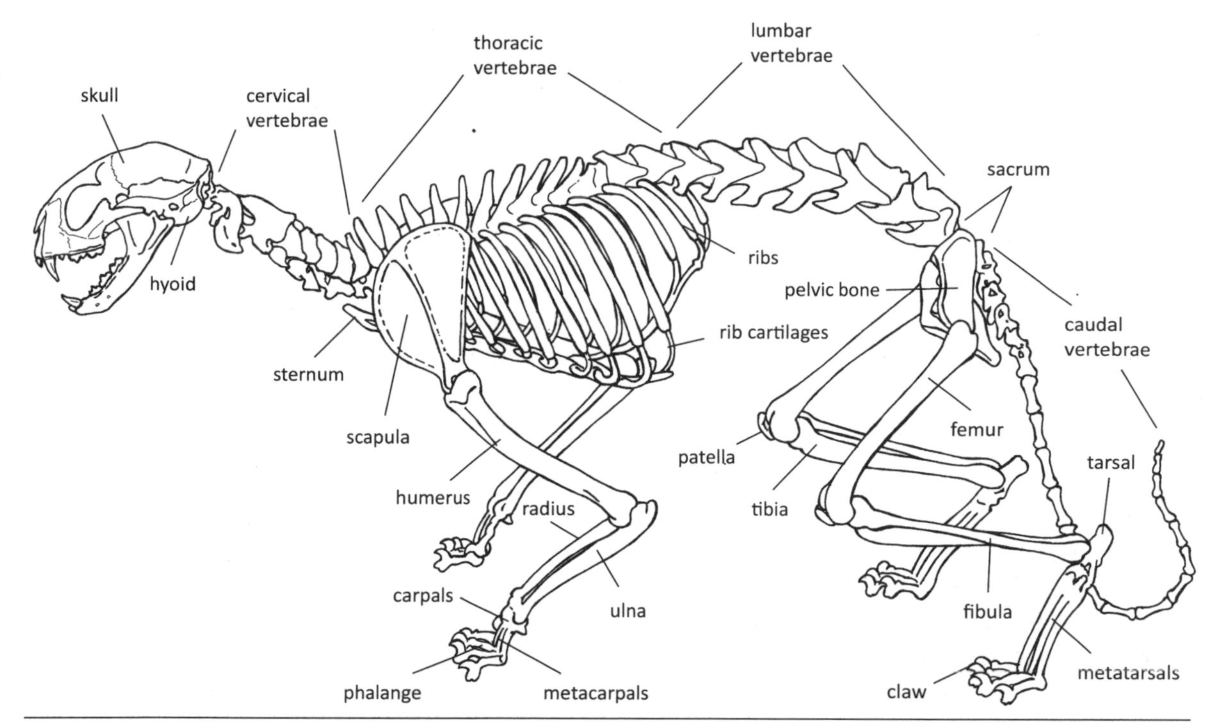

The cat's skeleton is made up of 244 individual bones, approximately 40 more than a human skeleton. This lightweight jointed framework is built around a central girder, the flexible spine to which the skull, rib cage, pelvic and pectoral girdle, and four limbs are attached. Unlike those in humans, the cat's scapula or shoulder blades are unattached to the skeleton. This allows superb flexibility, particularly at speed. See chapter 4 for a drawing with details of the hyoid.

and lower jaws. The second premolar is larger than the first, has two fangs securing it to the jawbone, and has a single flattened crown.

Often, domestic cats appear to be chewing with their rear teeth, but this is not so. A small cat's third premolars are highly specialized for cutting and shearing chunks of meat that are swallowed and not chewed. Unlike the flat, wide molars of a human that are used for grinding, the rear top and bottom teeth of small cats are used for sheering. These upper and lower premolars are called *carnassials* and slide against each other like the upper and lower blades of scissors. Unlike scissors, however, the upper and lower carnassials differ in shape. The upper carnassial is a large tooth with three fangs and one large crown, with smaller crowns on either side, while the lower carnassial is smaller and has two fangs and two crowns. Like self-sharpening scissors, the upper and lower carnassials on either side of the jaw wear against each other and remain sharp throughout the small cat's life.

Evolution has perfected the dentition of each small cat species over millions of years. Each species consumes different prey and so has a slightly different dentition. But evolution is even more subtle. The dentition is locally adapted, highly tuned to the kinds of prey a small cat species consumes.

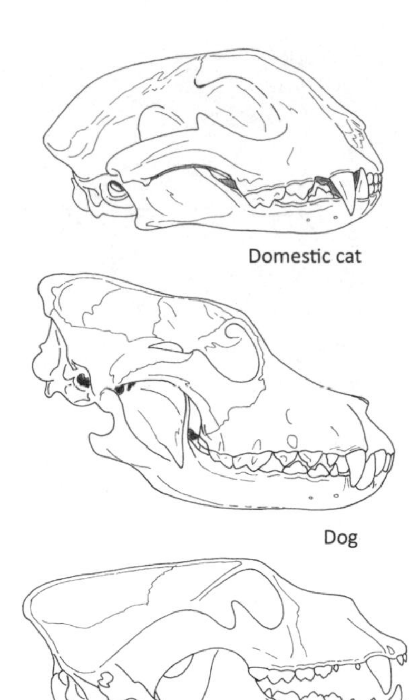

Cats have fewer but more specialist teeth than most mammals including humans. They have a far shorter jaw, less protuberant muzzle, and a more rounded head than dogs or bears. The hallmark of a cat's dentition is the carnassials, shearing teeth specialist for slicing meat.

This means that populations of the same small cat species living in different places with different prey are likely to have slightly different dentition. This simple fact causes great problems for taxonomists that use dentition and other morphological characters to classify species.

The skull contains the upper jaw, and the lower jaw attaches to the underside rear of the skull. The function of the jaws is to secure and kill prey and to tear pieces of meat and bone that are small enough to swallow. Function determines the form the jaws and skull take. Unlike many other animals, the jaws of a small cat operate quite simply: it only opens and closes. The jaws do not move side to side to grind food, because small cats don't grind food, and hence there are no grinding teeth. The jaw does not move forward or backward because it must lock tightly to prevent prey from escaping. The musculature that operates the jaw is anchored to the skull and back of the neck.

Other parts of the skull contain the sense organs. In some small cats, whose hearing is exceptionally well developed, the *auditory bullae* are prominent globular bulbs on either side of the underpart of the skull. In others, whose dentition more closely resembles that of a fish-eating otter, the front of the skull from the forehead to the snout is surprisingly flat, just like an otter's skull.

At 5 cm (2 in) long, clouded leopards have the longest canine teeth of all living cats and are about the same size as those of the average tiger. They are also capable of opening their jaws wider than any other cat.
Photo © Lon Grassman

The cat's open mouth reveals the four tusk-like canine or stabbing teeth. The "carnassials," or flesh-tearing teeth, are located in both the upper and lower jaws and are highly developed, extremely sharp, and capable of slicing through skin, soft bone, or cartilage. The solitary upper molar is rudimentary, while the incisors positioned between the canines are relatively small.

Front and rear paws. Not only do the front and rear paws of small cats closely resemble their larger cousins, they also resemble our own hands and feet. One obvious similarity is that small cats have five toes on their front or forepaws. The first toe of a small cat, the *dew-claw*, is shorter than the other toes and does not touch the ground. The dew-claw is analogous to the human thumb. One obvious difference is that small cats have just four toes on the rear or hind paws.

Small cats use their front paws to catch prey, climb trees, and defend themselves. While catching prey or climbing, the four toes of the front paws spread widely apart, just as does the human hand. The rear paws, like human feet, are no less important but are not used in the same fashion as the front paws. Just like human feet, the toes do not spread as far apart. This is best seen when observing small cat tracks in soft mud or wet sand. The front paws create the wider tracks, and the rear paws make the somewhat narrower tracks.

As with human fingers and toes, a small cat's toes have nails. Unlike a human's nails, small cat's nails are, in most species, fully retractable, that is, the nails rest protected and sharp within a sheath of skin. These retractable nails are one of the most notable attributes that separate wild and domestic cats from wild and domestic dogs. When a dog walks across the hard floor, its nails often make a clicking sound. Because its nails are retracted, a domestic cat makes no sound when walking across the same hard floor. As every cat owner knows, when needed these lethal weapons can instantly be summoned into action and swiftly wielded.

Tails. Much can be learned about a small cat's habits just from the tail. For instance, the Andean cat lives in the high, treeless, cold Andes of South America. The tail of this cat is extraordinarily large and bushy but feather light and blows in the slightest breeze. It is used as a scarf to cover the paws and nose when this cat is resting or sleeping in the bitter cold of the high Andes. Small cats that are highly arboreal, such as the margay of South America and the marbled cat of Southeast Asia, have fairly heavy long tails. Presumably, such tails act as counterweights to aid balance and help these cats move deftly through the trees. Small cats, such as the fishing cat and flat-headed cat, live on fish and frogs; their tails are short but muscular and help them swim when in pursuit of prey. Because form and function go hand in hand, knowing something about how a small cat functions enables accurate prediction of its tail's form; knowing details of the tail enables predicting a small cat's lifestyle.

Senses. Small cats share with all other animals and humans the five senses—sight, smell, sound, taste, and touch—but each is developed to a different degree in different species. All small cats have whiskers that appear in various rows across the face: above the eyes, in parallel rows across the muzzle, below the chin, and from the wrist. Unlike human facial hair, a cat's whiskers, called *vibrissae*, are extremely sensitive to movement caused by air currents or squirming prey. The vibrissae do not need to be in direct contact to sense an object. By walking with the vibrissae extending forward a blind cat can sense objects in its path by sensing air currents flowing around the object. So sensitive are cats to their external environment that many an owner of an aged domestic cat has been unaware that their house companion has been blind for quite some time.

The longer hairs extending from the small cat's wrists are also vibrissae. These hairs allow a cat to place its paw precisely without looking. Often, a domestic cat seems to place its paw tentatively, suddenly jerking it backward for some unknown reason. This is because the vibrissae are signaling information to the brain regarding where the paw is about to be placed. Wrist vibrissae enable a cat to walk silently across nearly any surface, and to stalk prey on the most difficult substrates without making a sound that would alert the unsuspecting prey.

Some small cats such as the serval, which live in African savannas, have large ears and long legs. Once again function has determined form. The serval, a rodent specialist, hunts prey that lives under cover of dense and often high grass. Large ears and acute vibrissae enable the serval to sense the slightest sounds beneath the grass; long legs enable a final aerial assault.

Evolution has produced a wondrously sensitive predator that carries no superfluous evolutionary baggage. Even a cat's purr has a vital, often lifesaving, function. The small cat we see today, wild or domestic, is nature's most perfect terrestrial predator and in action is the very essence of poetry in motion.

What is the difference between small cats and big cats?

Body weights of the wild cats range continuously from the smallest cat, weighing around 1.5 kg (3.3 lbs), to large cats weighing more than 100 times this weight. Male cats typically weigh more than the females. Deciding which wild cat species are "small" and which are "big" is often in the eye of the beholder. This task has been made much simpler due to human fascination with large predators. Indeed, most people are probably familiar with the seven big cats: the lion (Felidae *Panthera leo*), tiger (F. *P. tigris*), jaguar (F. *P. onca*), leopard (F. *P. pardus*), snow leopard (F. *P. uncia*), puma (F. *Puma concolor*), and cheetah (F. *Acinonyx jubatus*).

The phrase "small cats are big cats writ small" does not do the small cats justice. In fact, Felidae may be described as exhibiting biodiversity due only to the great diversity of small cats. Just 7 of the 37 members belonging to the Felidae are considered big cats. Because these seven are widely recognized large, extraordinary creatures, they are often referred to as *charismatic megafauna*. Because they are large, they are mostly terrestrial two-dimensional land predators.

Small cats occupy a greater range of habitats and space than big cats. Some small cats eat more aquatic prey and so, more often than not, stalk

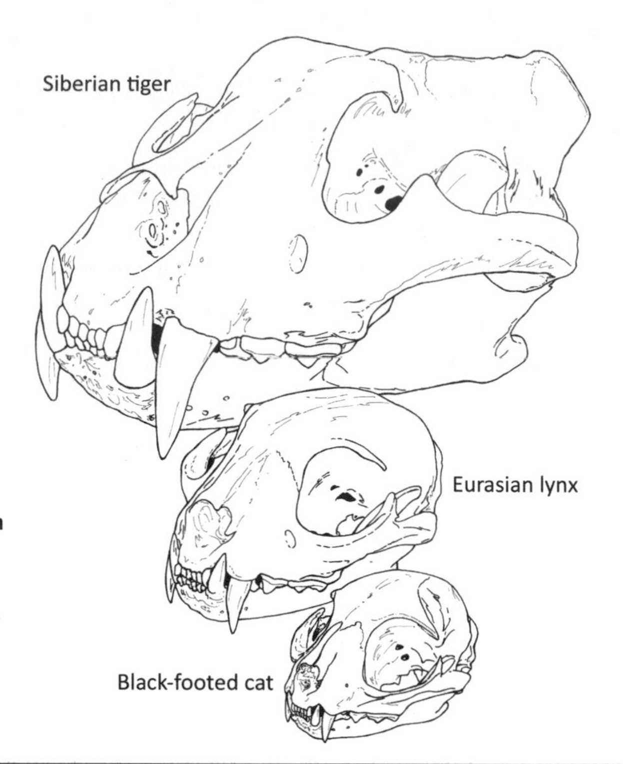

The differences between small and big cats include the skull size and shape. Because small cats prey upon smaller animals than do large cats, their muscle attachments are not as large. The tiger's skull is more than double the size of the lynx's, which in turn is over one-third larger than the small black-footed cat.

their prey in or near water. Other small cats are highly arboreal and make a living stalking their prey in trees; these cats have a three-dimensional lifestyle. At least one small cat lives in the desert sands of North Africa and the Middle East, while another makes its home in the boreal forest, habitats unoccupied by large cats.

Can this additional diversity be quantified? The answer is yes. Being at the top of the food chain means that predators are naturally rarer than their prey. Most predators defend the place where they live against all that would live in close proximity, especially other predators. Top predators can co-occur when they naturally avoid each other. This happens when each different species consumes different prey, are active at a different time of day, or occupy different habitats. Because some small cats are active primarily during the day (diurnal) or night (nocturnal), consume different prey, or are arboreal (live in trees), they co-occur with big cats and with other small cat species. These features have profound implications.

For instance, there are now 37 living species of wild dogs such as foxes, jackals, and wolves. Because wild dogs are strictly terrestrial, the species do not differ greatly in size (there is no wild dog as big as a lion or tiger), and because many are social, fewer species can co-occur. The maximum number of wild dog species known to co-occur is six, and this is observed in just one locale, southern Africa. Fully eight different wild cats species are known to co-occur in Southeast Asia. Of these eight species of wild cats, six are small cats occupying different habitats within the area occupied by two big cats. Even without the big cats, having six different small cats in a general area is unusual. We might, therefore, refer to the small members of the Felidae as *charismatic minifauna*, for they too are truly extraordinary creatures.

How many kinds of small cats are there?

Scientists now recognize 36 species of wild cats. The number increases to 37 if the domestic cat is included as a separate species. Of these, 30 are small cats. Since 1758, when Carl von Linné (Linnaeus) assigned Latin names to the lion, tiger, leopard, jaguar, ocelot, domestic cat, and Eurasian lynx, the number of species of wild cats recognized by scientists has ranged from about 20 to more than 50. However, the number of wild cats in nature has remained constant for at least the past 200,000 years.

This confusion among scientists arose because new wild cats were still being discovered in the late 1800s, and some species had multiple scientific names. Each new discovery increased the number of wild cat species, and because few specimens were available, the same species discovered in different places led to the naming of a new species, even though the species had been previously found and named. Confusion arose regarding the appearance of many wild cat species. Very similar individual wild cats, with different spot patterns and coat colors, were believed to be different species.

It was not until the late 1800s and early 1900s that the last new wild cats were discovered and named, and scientists realized that the same wild cat species could exhibit very different spot patterns and coat colors. In fact, we need only look at all the different coat patterns and colors of domestic cats; even those from the same litter often have different coat colors and patterns.

The placement of many species of wild cats in the Felidae family tree continued to confound the very best scientists into the late 1980s. Deciding if a wild animal is a wild cat is fairly easy, even for the casual observer, but deciding which species a particular wild cat is can prove a very frustrating task. Identification of some species is extremely difficult and continues to confound wild cat authorities, for instance, cats may show variation in tail length, coat color (from light gray to deep chestnut brown), spot patterns (that may differ on each side of the body), and body weight (which increases with increasing distance from the equator—the weight is proportional to the latitude).

Why are small cats important?

Small wild cats play a vital role in nature's economy. No population of any living biological organism can expand to conquer the earth. All biological organisms' populations are limited by some combination of predation, disease, and famine. As top predators, small cats eat rodents, birds, and just about any other wildlife they can catch. Generally, wherever small cats live, they prey upon what is most common or easiest to catch. In this way, top predators increase biodiversity by limiting the population of the most successful organisms, enabling other organisms to successfully compete with stronger competitors. No overly successful organisms can monopolize resources to the exclusion of others when a top predator preys upon them. Small cats acting as predators maintain higher biodiversity where they occur; in their absence, biodiversity is generally lower.

Small cats benefit humans. Many are highly skilled, efficient rodent predators. Rodents are pests of human food crops and also spread disease. For humans, a small cat that preys upon rodents should be a welcomed guest. Even Linnaeus recognized the value of a domestic cat when he referred to them as "the lion of mice."

Where do small cats live?

Small cats occur in an extraordinarily diverse range of habitats, from deserts to tropical forests, from grasslands of Argentina and Brazil to the arboreal forests of the Northern Hemisphere, and from deserts and swamps at sea level to more than 5500 m (18,000 ft) in the Andes of South America and the Tibetan plateau. Small cats occupy far more diverse habitats than big cats. As with many species, they occur naturally on all continents except Antarctica and Australia. They do not naturally occur on oceanic islands such as the Galápagos, Hawaii, Madagascar, and New Guinea. Small cats are found on continental islands off the coast of South America such as Trinidad, Chiloé Island off the coast of Chile, the Southeast Asian islands of Sumatra, Java, Borneo, Palawan, and three other islands that make up the Philippines. One small cat, the leopard cat, is found on Bali, but eastward across Wallace's Line, an imaginary line that passes between the islands of Bali and Lombok, no small cats are found. No small cats occupy the islands of Lombok, Sulawesi, or New Guinea.

Wherever there is sufficient prey, there is a good chance small cats are present. However, not all habitats are equally occupied. For instance, the sand cat lives in the deserts of North Africa and the Middle East but has no analog in the Western Hemisphere. The flat-headed cat found in the Malay Peninsula, Sumatra, and Borneo lives in proximity of swamps and

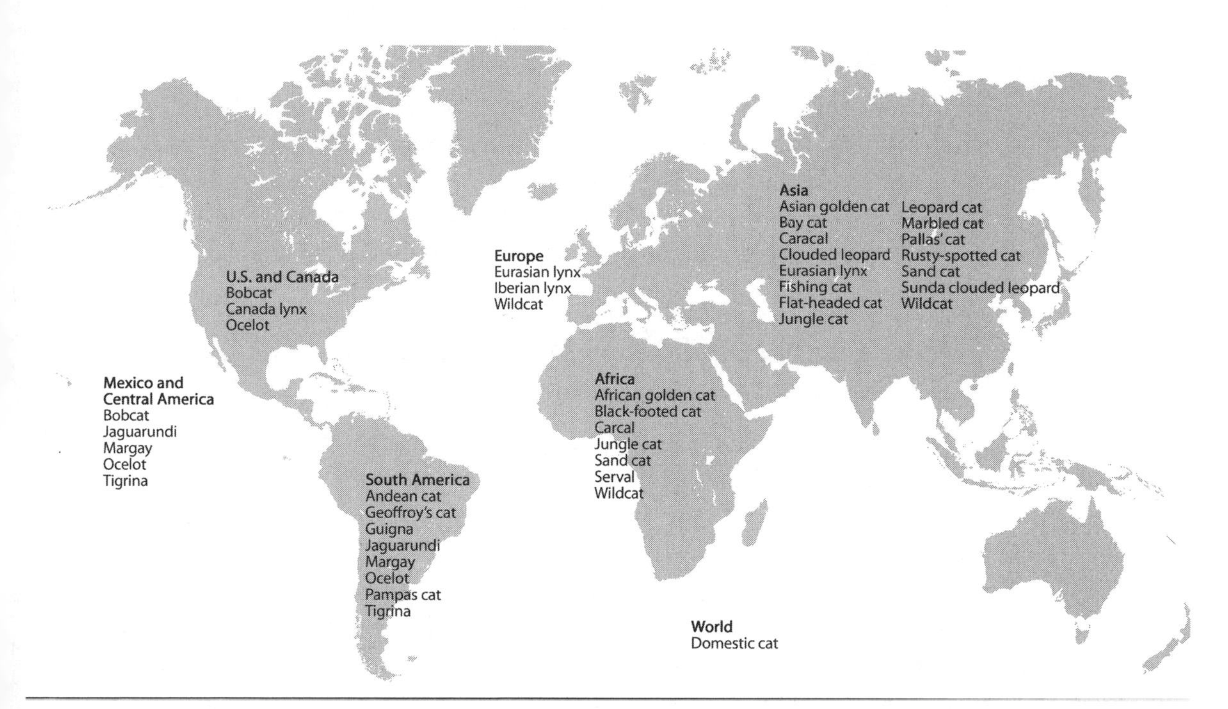

A world map showing the distribution of small cats. Note that some species of cat are found in more than one continent or geographic area. For example, the wildcat can be found in three different locations, Europe, Africa, and Asia. The domestic cat is the only species of small cat found globally.

streams, preying upon frogs and fish, but has no analog in Central or South America or Africa. However, members of the lynx genus are found in forests across the Northern Hemisphere.

Though there is an approximate one-to-one similarity between habitats on different continents, in many cases there is no one-to-one match between the small cats species found in these habitats. For instance, the black-footed cat of southern Africa and the pampas cat of South America, which live in savannas, might be considered analogs. The guigna of the temperate forests in southern Chile and Argentina might be analogous with the leopard cat found in temperate forests in China. The arboreal margay lives in trees in the tropical forests of Central and South America and so might be analogous to the arboreal marbled cat of Southeast Asia. However, there is no analog in Africa or the Western Hemisphere of the fish- and frog-eating flat-headed cat of Southeast Asia.

Could small cats live in Australia, Madagascar, or on oceanic islands where they do not occur naturally? The answer is yes, provided there is sufficient prey and the match is made between the small cat species and the most suitable habitat. From a small cat's perspective, prey is prey, regardless of its Latin name.

The complex geographic distribution of the world's small cats is best understood from an evolutionary perspective, for modern small cats are the product of millions of years of evolution. This process was well under way when the land bridge between North America and South America joined the two great continents and made possible the invasion of modern small cats into South America. Without an evolutionary perspective, it would be extremely difficult to explain the modern geographic distribution of small cats.

Why are there only domestic small cats on Madagascar, Australia, New Guinea, and New Zealand?

The current fossil record, although sparse, shows that the Felidae first appeared in Asia around 35 million years ago during the Oligocene, a geologically active time in the earth's history. Although this seems like a very long time ago, it is relatively late in the earth's geologic history and explains why cats appear where they are found today.

Long before the appearance of the first wild cats, Australia and its surrounding islands of New Zealand and New Guinea had separated from the supercontinent of Gondwanaland. The mammals occupying Australia have their origin predating the origin of the Felidae by more than 200 million years. There has never been a land bridge from Africa or Asia to Australia and its surrounding islands, even during times of low sea level. Thus, wild cats and other species, such as the primates, were never able to colonize the Australian continent. New Zealand and New Guinea were once part of Australia and also were never colonized by wild cats.

Africa was also a separate continent without cats. The geological record shows that Madagascar and also the Indian subcontinent were once connected to the east coast of Africa. Around 180 million years ago Madagascar and India separated from Africa. This was well before wild cats first appeared. The island of Madagascar and the Indian subcontinent also broke apart and India moved northward, slamming into the southern coast of the Asian continent. As part of Asia, India was eventually populated with early Asian wild cats. Africa was populated with wild cats when its northeastern coast became connected to SouthWest Asia.

Wild cats of Asian stock populated the islands off the coast of Southeast Asia, such as Java, Sumatra, Borneo, and Bali, during periods of low sea levels. The Indonesian island of Bali is separated from its eastern neighboring island of Lombok by less than 100 km. Tigers once occurred and leopard cats are presently found on Bali, but no wild cats are found on Lombok or its eastern neighboring island of Sulawesi, for Lombok and Sulawesi were once part of the great Australian land mass.

The naturalist and co-discoverer of evolution, Alfred Russel Wallace, a contemporary of Charles Darwin, explored some of the islands of Southeast

Asia. Wallace first called the world's attention to the differences in plants and animals between the islands of Bali (then with two wild cats species) and Lombok and islands east and south with no wild cat species. The imaginary line separating Bali and Lombok is called Wallace's Line in his honor.

Today, the small domestic cat has reached most of these continental islands and many oceanic islands as well, often followed by unintended consequences. The continent of Australia has also become populated with domestic cats that have had a devastating impact on native wildlife.

How are small cats classified?

Classification is simply a way of organizing or grouping complex information. In its simplest form, classification is a memory aid. To classify all living organisms, scientists use a hierarchical classification system that resembles a tree. The trunk of the tree represents all living species. A large branch might represent the class of mammals. From this large branch slightly smaller branches would represent each order of mammals: carnivores, primates, and so on. A branch of the tree protruding from the order of carnivores would then represent the Felidae, all the world's cats. From the Felidae branch would emanate eight smaller branches each representing a distinct lineage of cats. The leopard cat lineage represents all Asian small cat species. The leaves of the leopard cat branch represent the most closely related species. For instance, two adjacent leaves might represent two sister species: the flat-headed cat and fishing cat of Southeast Asia.

Small cats, like all cats, have been classified in two distinct ways. Determination of similarities and differences can be based upon morphological characters, or blood relatedness. The two ways to classify the small cats are not the same and do not produce the same classification of living small cats.

Before molecular analysis, the analysis of sequences of DNA, became popular, small cats, like all living things, were classified using morphological (physical) features such as the teeth and skull. Such a classification system places closely together those species that most resemble one another physically. For instance, if the teeth and skull of two unknown species are similar but not identical, then the two unknown species might be sister species and not the same species. For coarse-scale classification, the use of morphological characters works well and can be used to create a taxonomy. However, if a taxonomy based on relatedness is desired, another methodological choice is required. This is because familial relationships are passed on from mother to daughter in the mitochondrial DNA. Though we might not look like our parents morphologically, the truth is in our mitochondrial DNA (our phylogeny).

Table 1.2. Felidae lineage, scientific name, common name, and geographic location

Lineage	Scientific name	Common name	Geographic location
Bay cat	*Catopuma temminckii*	Asiatic golden cat	Southeast Asia
	Catopuma badia	bay cat	Southeast Asia
	Pardofelis marmorata	marbled cat	Southeast Asia
Caracal	*Caracal aurata*	African golden cat	Central Africa
	Caracal caracal	caracal	Africa, Middle East
	Leptailurus serval	serval	Africa
Felis	*Felis nigripes*	black-footed cat	Southern Africa
	Felis catus	domestic cat	global
	Felis chaus	jungle cat	Nile delta, Middle East
	Felis margarita	sand cat	North Africa, Middle East
	Felis silvestris	wildcat	Europe, Africa, Asia
Leopard cat	*Prionailurus viverrinus*	fishing cat	Southeast Asia
	Prionailurus planiceps	flat-headed cat	Southeast Asia
	Prionailurus bengalensis	leopard cat	Asia
	Otocolobus manul	Pallas' cat	Central Asia
	Prionailurus rubiginosus	rusty-spotted cat	India, Sri Lanka
Lynx	*Lynx rufus*	bobcat	North America
	Lynx canadensis	Canada lynx	North America
	Lynx lynx	Eurasian lynx	Eurasia
	Lynx pardinus	Iberian lynx	Spain (Iberian peninsula)
Ocelot	*Leopardus jacobita*	Andean cat	South America
	Leopardus geoffroyi	Geoffroy's cat	South America
	Leopardus guigna	guigna	South America
	Leopardus wiedii	margay	Central and South America
	Leopardus pardalis	ocelot	North, Central, and South America
	Leopardus colocolo	pampas cat	South America
	Leopardus tigrinus	tigrina	Central and South America
Panthera	*Neofelis nebulosa*	clouded leopard	Southeast Asia and Southern China
	*Panthera onca**	jaguar	Central and South America
	*Panthera pardus**	leopard	Africa and Asia
	*Panthera leo**	lion	Africa and Gir Forest, India
	*Panthera uncia**	snow leopard	Central Asia (high mountain regions)
	Neofelis diardi	Sunda clouded leopard	Borneo and Sumatra
	*Panthera tigris**	tiger	Asia
Puma	*Acinonyx jubatus**	cheetah	Africa and Middle East
	Puma yagouaroundi	jaguarundi	Central and South America
	*Puma concolor**	puma	North, Central, and South America

*Big cats

Since 1990, advances in molecular analysis have come at an increasingly rapid pace. These technological advances made possible analysis of familial relationships within the Felidae. The results of both classification schemes have been compared and differ greatly.

Modern molecular analysis done by Stephen O'Brien and his colleagues at the National Institute of Health in Fredericksburg, Maryland, shows that all 36 living wild cats and the domestic cat appeared in the last 11 million years from an ancestral wild cat that itself was the result of some 25 million years of evolution. All living wild cats can be placed in eight distinct lineages. Each of these lineages has unique characteristics, making each memorable. Understanding these lineages is the key to understanding all wild cats. Small cats are best understood using these lineages for two reasons: (1) small cats are found in each of the eight major wild cat lineages and (2) familial relationships aid our understanding of the distribution of modern small cats.

Let's begin with the easiest lineage. The panthera lineage includes some of the most fearsome and most widely recognized terrestrial mammalian predators on earth. Because all cats are members of the Felidae, the family name Felidae will henceforth not be included when using the Latin designation.

The panthera lineage includes five big cats and two small cats. The five largest members, the lion, tiger, leopard, snow leopard, and jaguar, are all closely related and so placed in the genus *Panthera*. The panthera lineage also includes two small cats: the clouded leopard and its sister species, the Sunda clouded leopard, that split very early from the rest of the lineage and are now placed in their own genus, *Neofelis*.

The puma lineage contains the remaining two big cats, the puma (*Puma concolor*) and cheetah (*Acinonyx jubatus*), which are closely related. The puma lineage also includes the small jaguarundi (*Puma yagouaroundi*). These three species share the common trait of having relatively small heads compared to their bodies. Because the cheetah evolved early in this lineage it is placed in its own genus, *Acinonyx*, while the jaguarundi and puma are closely related American species. Molecular evidence suggests that the cheetah evolved in North America and migrated into Asia during a period of low sea level, and from there into Africa. Unlike the cheetah, adult jaguarundi and puma lack spots. Indeed, *concolor* is Latin meaning "of a uniform color."

The bay cat lineage of Southeast Asia has three representatives, and all are small cats: the Asiatic golden cat (*Catopuma temminckii*), the bay cat (*C. badia*), and the marbled cat (*Pardofelis marmorata*). The bay cat, found only on the island of Borneo (we say the bay cat is *endemic* to Borneo), looks like a three-quarter-sized version of the Asiatic golden cat, which is itself about one-half the size of a puma. The bay cat lineage contains two of the most poorly known wild cats: the bay cat and the marbled cat.

The lynx lineage contains all four species of lynx. All are found only in the Northern Hemisphere: the bobcat (*Lynx rufus*), Canada lynx (*L. canadensis*), Eurasian lynx (*L. lynx*), and Iberian lynx (*L. pardinus*). All are small cats that share several common traits: a short tail, large paws, and long, pointed hairs on the ends of their ears. The bobcat and Canada lynx are found in North America.

Seven other small wild cats found in the Americas are members of the ocelot lineage. The largest member is the ocelot (*Leopardus pardalis*), and the remaining six are the margay (*L. wiedii*), tigrina (*L. tigrinus*), Geoffroy's cat (*L. geoffroyi*), pampas cat (*L. colocolo*), guigna (*L. guigna*), and Andean cat (*L. jacobita*). All wild cats have 19 pairs of chromosomes except those in the genus *Leopardus*, which have 18 pairs of chromosomes. This trait leaves little doubt that members of the ocelot lineage evolved from a common ancestor.

The leopard cat lineage of Asian origin contains five small cats: Pallas' cat (*Otocolobus manul*), flat-headed cat (*Prionailurus planiceps*), fishing cat (*P. viverrinus*), leopard cat (*P. bengalensis*), and rusty-spotted cat (*P. rubiginosus*). The flat-headed cat and fishing cat are the most aquatic of all wild cats.

Although not strictly African, two of the three small cat members of the caracal lineage are found only in Africa: the African golden cat (*Caracal aurata*) and the serval (*Leptailurus serval*). The caracal (*C. caracal*) is found from Africa to India. In a family of exceptionally gifted jumpers, the caracal and serval are standouts.

The felis lineage that contains the domestic cat (*Felis catus*) is also of Asian origin, and the other four small members are found in Eurasia and Africa. The jungle (or swamp) cat (*F. chaus*), sand cat (*F. margarita*), black-footed cat (*F. nigripes*), and wildcat (*F. silvestris*). All members of the felis lineage are closely related to the domestic cat.

What characterizes the major groups of small cats?

Each of the eight lineages contains small cats: caracal lineage, puma lineage, ocelot lineage, bay cat lineage, lynx lineage, leopard cat lineage, and felis lineage. Each lineage can be thought of as a branch off the main stem of the cat family tree. Each member of a lineage shares unique gene sequences and morphological characters and geographic locations. Since we cannot see gene sequences, in most instances morphological characters can be used to distinguish the eight lineages. For instance, two of three small cat members in the caracal lineage are found strictly in Africa. Two of the members of this lineage are the great leapers, the high-jumpers, of the cat world.

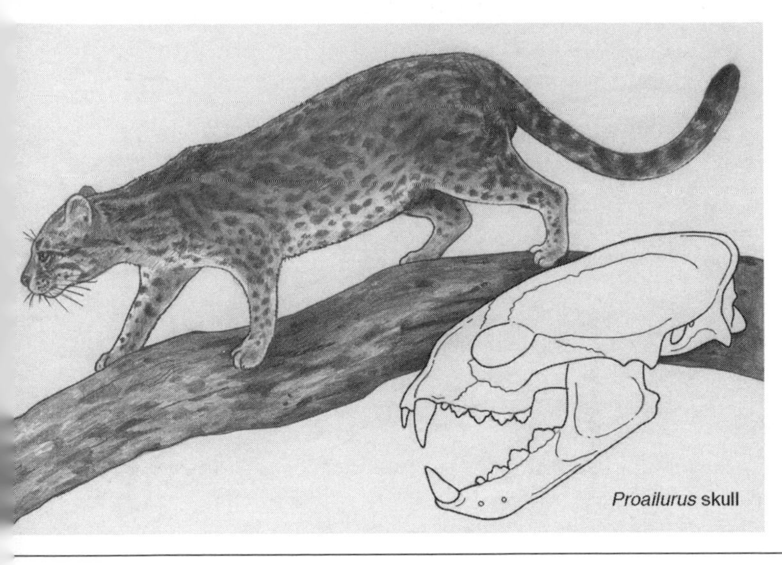

An artist's impression of *Proailurus*, the extinct "first cat," using scientific and fossil evidence. It was believed to have been a weasel-like cat about the size of an ocelot, a skilled climber with an arboreal lifestyle, and with an extensively marked sandy or light ochre coat.

The puma lineage is unique among the eight lineages for two reasons. First, two of its three members are big cats—the puma and the cheetah—and only one is a small cat—the jaguarundi—that is perhaps the most un-cat-like member of the Felidae. Second, all three members share a common trait found only in this lineage—each member has a small head compared to its body size. All three members also have long tails, but this trait shows up in other lineages as well.

All seven members of the ocelot lineage are found in Central or South America. This is perhaps the most confusing lineage because many of the members have similar coat colorations, and one has a large variety of coat pattern and color differences. All are closely related and placed in the same genus, *Leopardus*.

The bay cat lineage has just three members and is arguably the most poorly known lineage. All the members of the bay cat lineage are small cats with long tails.

Without question all members of the lynx lineage have the shortest tails compared to their body size of all wild cats, large and small. Each member also has long hairs at the tips of each ear. All four members are placed in the same genus, *Lynx*.

The felis lineage also consists entirely of small cats. All members evolved relatively recently and so are placed in the same genus, *Felis*. With the exception of the jungle cat, members of this lineage most resemble the familiar member of many households, the domestic cat, which is also placed in this lineage.

The leopard cat lineage is of Asian origin and has five members, all small cats. The leopard cat lineage contains the two most aquatic members

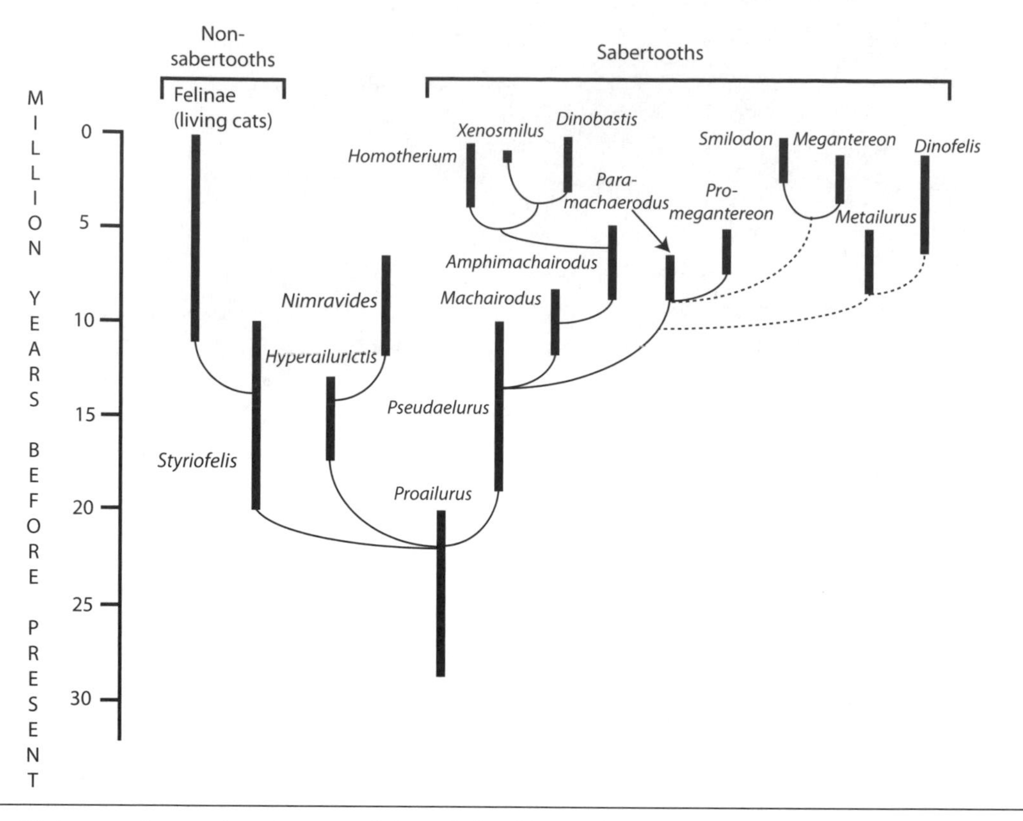

Felidae Phylogeny. A tentative diagram of the relationships within the cat family from its origins until today
Modified from L. Werdelin et al., 2010

of the Felidae. Although jaguars and tigers do not hesitate to enter water, two members of the leopard cat lineage are usually associated with water. While jaguars and tiger hesitate to put their heads underwater, the flat-headed cat and fishing cat show no such inhibition.

When did small cats evolve?

The most recent fossil evidence suggests that what can be considered true cats arose around 35 million years ago in the Oligocene. The first true cat is placed in the genus *Proailurus*. *Pro* means early and *ailurus* is a Latinized version of the Greek word, *ailouros*, meaning cat. The only known species was *P. lemanensis*, whose fossils have been found in France. This was a cat the size of an ocelot that differed mainly from all living cats in not having lost the front and hind cheek teeth. Because this first true cat's ancestors were not cats but more like civets, the dentition is more similar to its relatives than to living wild cats.

What is the oldest small cat fossil?

The oldest specimens of *P. lemanensis* date back about 20 million years and come from St.-Gérand-le-Puy in France. Incomplete specimens are known from the Quercy region of southern France in sediments that are about 30 to 27 million years old. Based on these fossils, scientists believe the first true cats appeared in Europe.

Wild cats might be rare, but fossil cats are exceedingly rare and consist mostly of fragments. Each new discovery adds to existing knowledge and often alters what was previously believed. Not surprisingly, the very earliest cats were cat-like in their features. Aside from the greater number of teeth that gave rise to a longer snout, early cats do not appear to differ in any appreciable way from living cats. Early cats had generalized skeletons and retractile claws, were likely skilled climbers, and had dentition that was adapted to cutting and consuming flesh. Since these early beginnings so very long ago, the family of cats has evolved conservatively by experimenting with and refining to the highest degree those characters that proved most useful. The absence of great changes in the form and function of small cats is in complete contrast to other carnivore families such as bears and hyenas.

Chapter 2

Form and Function of Small Cats

What are the largest and smallest small cats?

Although this appears to be a simple question, the answer shows otherwise. Just as in humans, body weights of individuals in each species of small cats vary. Generally, male small cats are larger than female cats. The difference in body weight can be as much as 30%. The weight of individuals within a species varies by habitat, elevation, distance from the equator, and even, with all other factors being equal, whether they are found on the mainland or an adjacent island. Variation is the norm, be it in color, tooth morphology, or body weight and size—for small cats that have spots, the spot patterns on either side of an individual's body differ! Even for the smallest small cats, size differences are sometimes striking.

The guigna might be the New World's smallest cat. Although the tigrina weighs as little as a guigna, the guigna is shorter from the tip of the tail to the tip of the nose. The guigna, a member of the ocelot lineage, is found in south-central Chile and a small part of south-central Argentina. Guignas are also found on some islands off the coast of Chile. One such island where the guigna occurs is Isla Grande de Chiloé, where JGS was the first person to study the species in the wild. On the mainland, female guignas weigh around 1.4 kg (3 lbs) and males weigh in at a hefty 1.7 kg (3.75 lbs). This is much smaller than a typical domestic cat. At the same latitude and altitude as they occur on the mainland, male and female guignas on Isla Grande de Chiloé weigh 2.2 kg (4.8 lbs) and 1.7 kg (3.75 lbs), respectively. Females on the island are as big as mainland males. Guignas on Isla Grande de Chiloé exhibit what is referred to as *island gigantism*, that is, guignas are larger on islands than on the adjacent mainland.

Table 2.1. Small cat size comparison

Species	Body size (kg)	Head and body (cm)	Tail (cm)
clouded leopard	11–22	82–96	60–85
Sunda clouded leopard	11–22	82–96	60–85
Eurasian lynx	12–25	80–140	11–24
caracal	11–20	60–105	20–35
bobcat	7–18	65–100	13–15
Canada lynx	8.5–16	73–100	10–13
Asiatic golden cat	12–16	50–80	35–56
Iberian lynx	9–16	71–98	10–13
serval	8–16	67–100	24–45
jungle cat	4–16	50–90	20–31
fishing cat	5–15	65–100	24–41
ocelot	7–14	66–100	26–41
African golden cat	5–14	61–102	16–37
wildcat	3–8	45–75	21–35
leopard cat	2–7	44–95	15–44
jaguarundi	3–6.5	51–77	28–51
bay cat	3–6	50–67	32–40
pampas cat	3–6.4	50–70	22–33
Geoffroy's cat	2–6	43–70	23–37
marbled cat	3–6	45–62	35–55
Andean cat	3–5	57–65	41–48
Pallas' cat	2.5–4.5	56–65	21–31
margay	2.5–4	47–79	33–50
domestic cat	3–4	40–60	25–35
sand cat	2–3.4	40–57	27–35
guigna	1.4–3	39–51	20–25
tigrina	1.75–2.75	40–55	25–40
black-footed cat	1.5–2.75	33–50	15–22
flat-headed cat	1.6–2.2	42–55	13–20
rusty-spotted cat	1–1.6	35–48	12–25

If we use as our units of measure the average body weight of males and females, and the average head and tail length of a species, then the rusty-spotted cat is the world's smallest wild cat. The diminutive rusty-spotted cat found only in India and Sri Lanka tips the scales at no more than 1.6 kg (3.5 lbs) and from the tip of the nose to the end of the tail measures less than 60 cm (2 ft).

As for the largest small cat, there is also no clear winner. An average Eurasian lynx has a body weight of approximately 20 kg (44 lbs) and, despite its shortened tail, a length that can reach 1.4 m (4.6 ft). A large clouded leopard can weigh as much as an average Eurasian lynx, but because of its long tail, can exceed 1.8 m (6 ft) from the tip of the nose to the tip of the tail.

Three very different wild cats are illustrated: the diminutive rusty-spotted cat *(front left)*, the larger clouded leopard *(right)*, and the Siberian tiger *(rear)*. One can readily see just how large the tiger is compared to its smaller cousins. There is a considerable size difference even among the 30 small cat species.

The tigrina of northern South America is one of the world's smallest cats, weighing just 2.2 kg (5 lbs).

To make matters more confounding, if the tail length is excluded from the largest small cat contest, then the length of the Eurasian lynx head and body easily exceeds that of the clouded leopard and, in fact, also exceeds the head and body length of three big cats with generously long tails—the cheetah, leopard, and snow leopard.

For our purposes, let's agree that the rusty-spotted cat found in Sri Lanka and India is the smallest cat in the Felidae. The clouded leopard of Southeast Asia is the largest small cat. The clouded leopard weighs 12 times as much and, from the tip of the nose to the tip of the tail, is three times as long as the rusty-spotted cat.

How fast does a small cat's heart beat?

Generally mammalian heart rates are inversely proportional to their body size. This means that the smallest small cats have a faster heart rate than that of the largest small cats.

In Chile, Constanza Napolitano holds an adult female guigna that has been lightly drugged for examination. Guigna females are very small and weigh less than 2 kg (4.4 lbs). Guigna males are heavier.

A resting domestic cat's heart rate is normally between 140 and 160 beats per minute. A resting domestic cat normally has a respiration rate less than 40 breaths per minute. Since most small cats are about the same size as domestic cats, a heart rate of 150 beats per minute approximates that of most small cats. Larger small cats such as the Eurasian lynx and clouded leopard will have slower heart rates, on the order of 90 beats per minute. Conversely, the rusty-spotted cat, black-footed cat, tigrina, and guigna can be expected to have resting hearts rates of around 180 beats per minute.

The heart and respiratory rates of most small wild cats is unknown. This is because most are sedated before they can be handled. The drug that is used causes the heart to beat abnormally, rendering heart and respiratory rate measurements inaccurate.

Heart rates also depend on the situation in which a small cat finds itself. Not surprisingly, a resting small cat's heart rate can be half that of the same individual fleeing a larger predator.

Recent studies suggest that mammalian heart rate is inversely propor-

The rusty-spotted cat of India and Sri Lanka is also one of the smallest of the world's cats and weighs less than 2.2 kg (5 lbs).

tional to lifespan. We know that elephants have a longer lifespan than a mouse, for instance. Thus, we should expect that larger small cats have a longer lifespan than smaller small cats.

Can small cats see color?

Yes, but not like humans can. Both small cats and humans have binocular vision. However, small cats, like most mammals except humans, are *dichromats* and have what is known as *dichromatic vision*. Normal humans are *trichromats*. Color-blind humans have dichromatic vision. Dichromats can, for example, differentiate red or green from other colors but cannot distinguish red from green. In humans, dichromatic vision is considered a defect; in small cats and most mammals dichromatic vision is the norm.

A mammal's eye is made up of light sensitive cells two of which are called *rods* and *cones*. Cones function well in bright light and detect color; rods lack the pigments found in the cones that are necessary to detect colors, but function well in low light and sense movement more quickly. Each contributes to what we call vision.

Being trichromats, humans have three different cones that detect short-, medium-, and long-wave light, sensitive to blue, green, and red light, re-

Despite its short tail, the Eurasian lynx is one of the largest small cats. In winter the coat is thick and heavy, making the cat appear larger.

The clouded leopard, is highly arboreal. Its name in Malay, "kuching dahan," translates literally as "cat fork-in-a-tree."

spectively. Being dichromats, small cats have cones that detect blue and green light, but lack the third for red light. In our human world, we would consider small cats color-blind. However, what small cats lack in cones is made up for in rods.

The number of rods and cones contained in the eye is space-limited, so having more rods necessarily means having fewer cones. Because those small cats that were successful night hunters reproduced more often, over time, evolutionary forces led to small cats having far more rods than cones. The evolutionary significance is clear: small cats hunt by day and night. Night hunting requires superior night vision and hence more rods.

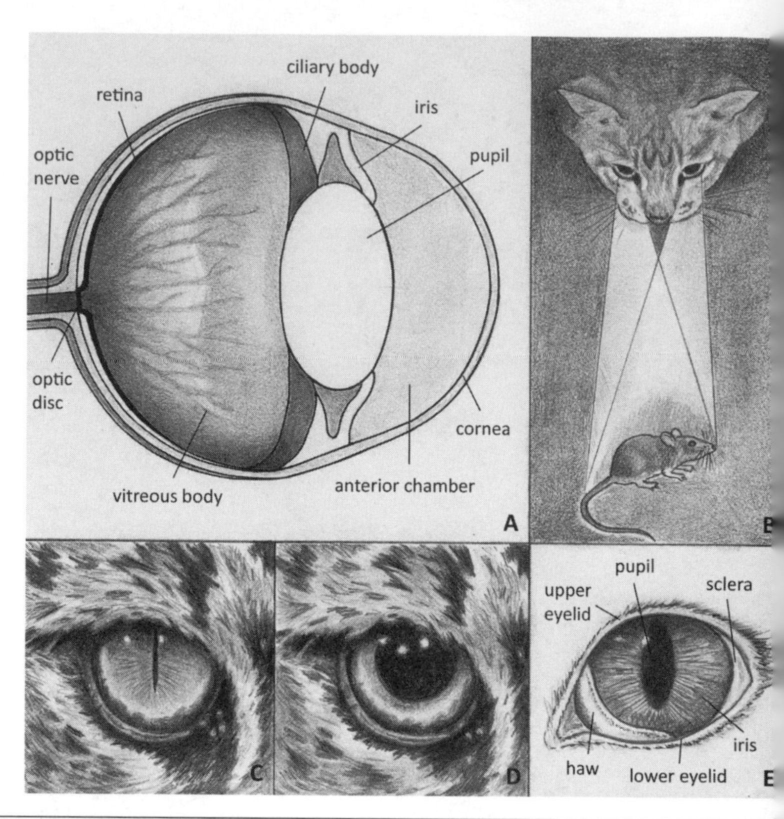

Cat's eyes differ from humans internally *(A)* and externally. Cat's eyes provide exceptional binocular vision *(B)*. The pupil is narrow in bright light *(C)* and large in low light *(D)*. Cats have an additional eyelid called the nictating membrane *(E)* that lubricates the corneal surface.

Distinguishing the colors of potential prey items under any light conditions is unnecessary. The increase in the number of rods reduced the possible number of cones, and the ability to see in color never was never fully developed. Thus, the night vision of small cats is notably superior to that of humans.

Much of what we know about the vision of small cats comes from studies of domestic cats. Most everyone has noticed that the eyes of domestic cats reflect light at night. This is referred to as *eye shine*. All small cats exhibit a similar night eye shine when light is directed on their eyes. Eye shine is caused by a special reflecting membrane called the *tapetum luvidum*, which reflects any light not absorbed by the rods during passage through the retina. The small cat's retina receives a second pass of light, thus increasing the small cat's ability to see in dim light. Not only do small cats have an increased number of rods, perhaps more than five times the number found in a human eye, they also have a mirror-like membrane that multiples the rod's ability to absorb light. Small cats would consider humans severely night-vision impaired!

Note that though it is often said that small cats can see in total darkness, this is not so. Some light must pass through the cat's eye before it to be absorbed and the signals processed by the brain. It's likely that other senses such as hearing and touch are relied upon in an environment devoid of all light.

While JGS was studying guignas in Chile, he observed with night-vision binoculars a female guigna from very close range on a moonless night under thick canopy of trees. She was an adult but less than half the size of a typical domestic cat. Crouched, her muscles tense, she was looking upward. Her head was quickly moving as if she was observing a quickly moving object just a few feet away. However, he could not see the object. Suddenly, she leaped upward, her paws swiped the air, and she landed on all four paws. Before she landed, the tiny moth she caught had been transferred to her mouth. Her ability to focus on a flying moth at night in very low light, the incredible acceleration she achieved, and her precision in the air was truly amazing and something no lion or tiger could ever master.

Are small cats capable of retracting their claws?

Yes. In fact, at rest the claws of most species of small cats are fully retracted. Small cats depend on their claws for climbing trees and rocks, securing and grasping prey, and as weapons of defense against aggressors. The importance of claws in a small cat's life cannot be overestimated. Evolution has produced a lethal weapon that when not in use is fully protected, enveloped within a fleshy sheath of skin. When necessary, contraction of tendons causes the terminal phalanx (last digit of the toe) that supports the claw to extend, thus exposing the claw. Another ligament functioning as a spring and attached to the claw becomes fully stretched when the claw is extended. When the tendons relax, the spring-like ligament tightens, withdrawing the claw into its protective sheath.

Can small cats swim?

Yes, all small cats are fully capable swimmers. However, being a capable swimmer does not imply small cats like water. Most small wild cats go out of their way to avoid water. They prefer to leap over small streams rather than wade in and walk across. However, if a stream cannot be jumped, small cats will wade in and swim to reach the opposite side. The flat-headed cat and the fishing cat regularly catch aquatic prey and often put their heads underwater to search for prey. Interestingly, the fishing cat has partially webbed front toes and a tail that is thick and muscular towards the tip, which some scientists believe might be used as a rudder. The tail of the flat-headed cat is also quite stumpy, but almost nothing is known of flat-headed cats in the wild. Guignas are also not shy of water.

Retractable, sharp claws are universal among small cats, but some retract their claws more than others. Cats use their claws to secure prey. Each front paw has five clawed toes *(left)*, and each rear paw has four clawed toes. Unlike dogs, cat's claws are normally retracted and kept within a protective sheath and are exposed when extended *(right)*.

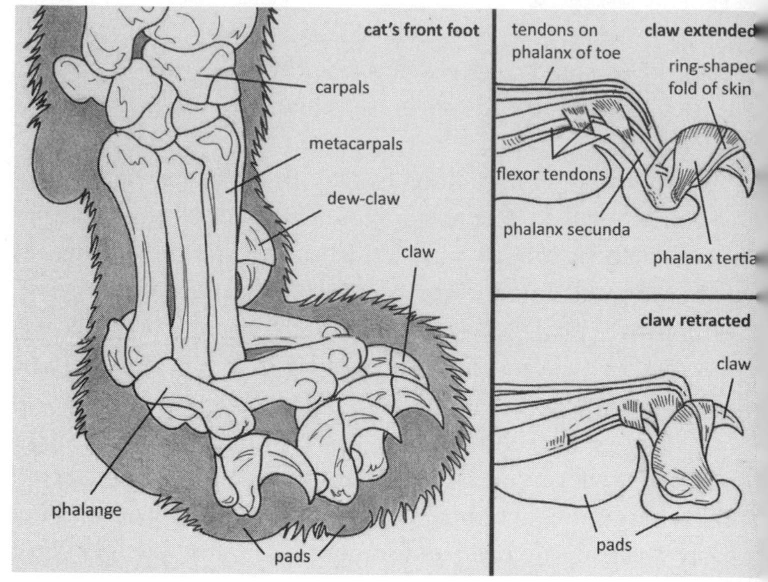

By gently squeezing on the toe, the claw of an ocelot is exposed. When released, the claw fully retracts and remains protected. Photo © Neville Buck

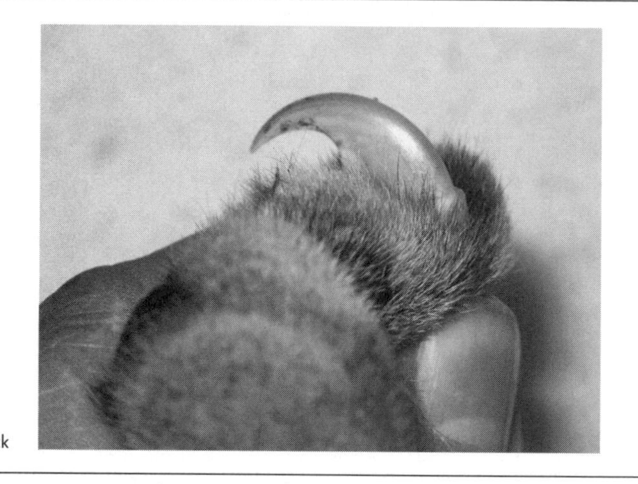

How far can small cats jump?

Although most small cats are capable of jumping, naturally some small cats jump higher than others. In fact, several are great leapers, as can readily be deduced from the relative length of their legs. The long-legged small cats like the caracal, serval, and jungle cat use their great jumping ability to secure their prey. The obvious way to tell if a small cat is a superior jumper is to judge how long the legs are compared to the body size and weight and the length of the tail. The caracal, serval, and jungle cat all have long limbs, relatively small tails, and slim bodies but are otherwise just as muscular as other small cats. No wild dog can out-jump these cats.

Caracals have been filmed in the wild jumping over a 3 m (9 ft) fence to prey upon sheep. In the 1800s in India, the jumping ability of tame caracals

A fishing cat attempts to secure a fish the easy way. Photo © Neville Buck

Failing to catch the fish with its paw, a fishing cat searches for it prey by submersing its head and looking underwater. Photo © Neville Buck and Jim Sanderson

was tested in contests to see how many pigeons could be knocked out of the air.

Servals have a remarkable way of hunting in tall grass. Unable to see their prey, the serval uses its large funnel-like ears to hunt by sound. Slowly walking through tall grass, the serval stops periodically to listen for slight movements. They may even wait for a number of minutes while they survey the area for the sound of a potential meal. Once prey is detected, the serval uses its large ears to locate the exact position of the sound. The serval then pounces high in the air and begins its aerial assault on its unsuspecting intended victim. If unsuccessful, the serval employs its unique hunting technique of executing a series of stiff-legged, bouncing jumps into the air,

The flat-headed cat, a smaller sister species of the fishing cat, also eats fish, frogs, and other aquatic prey.

The flat-headed cat also submerges its head underwater to secure aquatic prey. Both the fishing cat and flat-headed cat are considered endangered because of extensive, widespread conversion of their lowland habitat to oil palm plantations and loss of mangrove to shrimp farming.

A captive margay stalks a frozen bird attached to a tree. Note the orientation of the hind forelegs and paws that enable the margay to descend down a tree trunk head-first and in complete control. Margay can hang upside down from a single back paw, or run hanging upside down along a horizontal rope. Photo © Neville Buck

each time landing on one or both forepaws. Eventually, the prey is secured. Servals can easily leap 3 m (9 ft) into the air to catch birds on the wing.

Are small cats capable of climbing?

While all small cats are certainly capable of climbing, some small cats are better adapted to jumping, while others are better adapted to climbing. Indeed, most small cats readily climb to avoid predators or when otherwise threatened. Just as those small cats with long legs and short tails are best adapted at jumping, small cats with relatively long tails excel at climbing. Some exceptionally acrobatic species, such as the margay of Central and South America and the marbled cat of Southeast Asia, are highly arboreal and spend most of their time in the trees. The clouded leopard, Sunda clouded leopard, margay, and marbled cat also have wide forepaws that apparently aid their climbing ability.

An amazingly agile climber, the margay is a true acrobat and quite capable of performing incredible feats of aerial gymnastics. This cat is also capable of climbing down a tree truck head-first and hanging from a branch by its hind paws while using the front paws to grasp prey. Like a squirrel, the margay, marbled cat, and both clouded leopard species can rotate their hind paws enabling their ability to climb head-first down a tree. No other cats in the family Felidae share this unique ability. Both the margay and the marbled cat have slender bodies, broad feet, and long tails—obvious adaptations for an aerial lifestyle where balance is vital to survival.

Are small cats' tails the same?

The small cats we see today are the product of a long and complex evolutionary history. A small cat's morphology, or physical traits, has been perfected by millions of years of selection pressures. During this long history, those small cats that were better suited to their environment produced more offspring. In this way, favorable traits were passed on to offspring. Thus, every morphological feature, such as the tail, is not just a random appendage, but a fully functional completely tested part of a complete package, carefully adapted to making a living in the wild.

For example, the Andean cat and Pallas' cat have long, fluffy, and extremely lightweight tails. The Andean cat is supremely adapted to living in the high Andes that are extremely cold and windswept. Pallas' cat lives on the Asian steppe—another cold, dry, and often windy environment. Both cats' tails function like a beautifully fitted scarf capable of covering the nose and paws from the cold and bitter winds.

The most arboreal of small cats, the margay, both clouded leopards, and marbled cat have wonderfully long and comparatively muscular tails, adapted for improving balance. Anyone lucky enough to observe a margay in its natural state would be left in awe at the cat's magnificent aerial gymnastics. As the margay leaps about in the branches it's easy to see how well it uses its tail as the perfect counterweight.

In contrast, members of the lynx lineage, the bobcat, Canada lynx, Iberian lynx, and the Eurasian lynx, all have extremely short tails, but exceptionally wide paws, enabling the cats to pursue their prey in deep snow. We can only surmise that any other tail would be a disadvantage.

The Asiatic golden cat and the African golden cat have muscular, heavy tails. Apparently, these tails are used as counterweights during the pursuit of prey.

Do small cats have whiskers?

All small cats have about twenty-four movable whiskers. These whiskers appear in four or more parallel rows above the upper lip on each side

The shape of a cat's tail varies considerably between species. A bobcat's tail is extremely short *(A)*, a serval's tail is of medium length *(B)*, a clouded leopard's tail is long and heavy *(C)*, and Pallas' cat has a thick, bushy well-furred tail *(D)*.

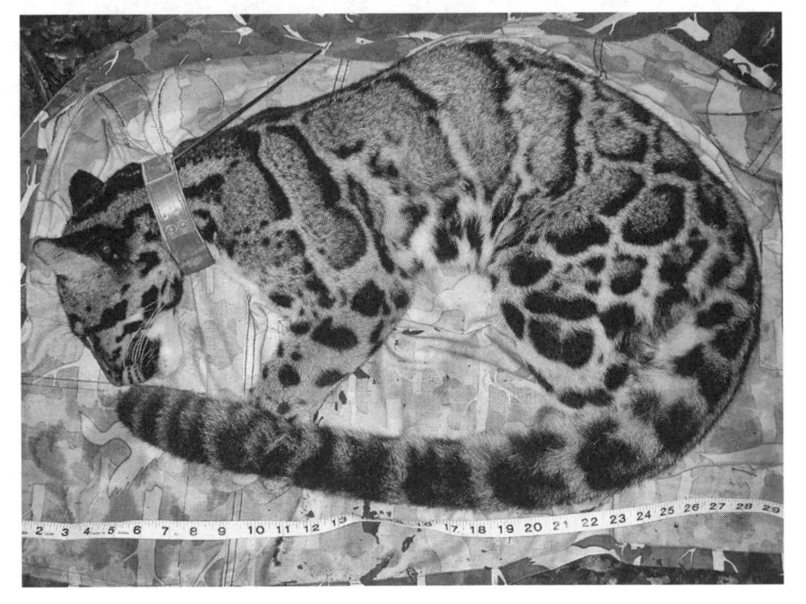

While sedated and radio collared, this clouded leopard shows off its extremely long tail and beautifully marked coat. The cat's tail is almost the same length as its head and body. Photo © Lon Grassman

The Andean cat lives in one of the coldest environments on earth, the high Andes of South America. Its tail is long and thick and yet is feather-light and blows in a slight breeze. The tail functions as a scarf that is draped over the paws and nose when the cat is sleeping.

The jaguarundi has a very long tail for a small cat.

of the cat's nose. However, whiskers are best considered as a special case of *vibrissae*, special sensory hairs.

There are vibrissae on each cheek, in tufts above the eyes on the forehead, on the chin, on the inner "wrists," and also at the back of the legs. Unlike human facial hair, vibrissae can be deliberately manipulated and pointed. For instance, the upper two rows of vibrissae on a small cat's face can move independently of the lower two rows. Vibrissae thus aid navigation and the sense of touch by picking up air currents flowing around an object, or, when they are in direct contact with an object, movement.

The bobcat is named for its short tail. Photo © Kenneth Adelman

The Chinese mountain cat is closely related to the domestic cat and has a similar set of whiskers across its face. Note also the pencils of hair protruding from the tips of the ears.

Vibrissae are thicker than ordinary hairs, and their roots are set deeper into the dermis in wells of fluid. When held facing forward, facial vibrissae provide information on objects that a small cat might not see, and enable walking and stalking in near total darkness. Even subtle changes, such as vibrations or changes in air currents, can be sensed.

The face of an ocelot shows multiple rows of vibrissae—sensitive whiskers. Photo © Neville Buck

Vibrissae are found in multiple places on a small cat's face. This ocelot has vibrissae on the chin below the mouth, multiple rows on the cheeks at either side of the nose, one group on each side of the rear jaw, and one group above each eye. Photo © Neville Buck

How sensitive is a small cat's sense of smell?

Not much is known about small cat's sense of smell. What has been learned comes almost entirely from studies of domestric cats. A domestic cat's sense of smell is estimated to be approximately 15 times more sensative than a human's sense of smell. This should not be surprising because humans depend far more on sight than on smell. Unlike humans, domestic

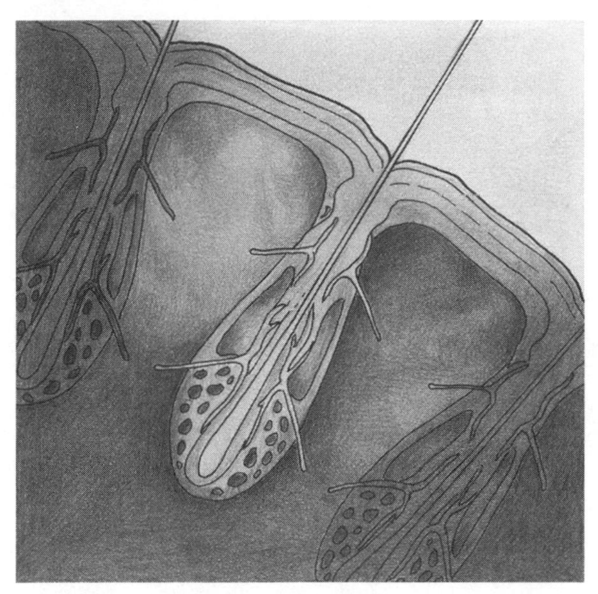

Whiskers, or vibrissae, are thicker than ordinary hair and are found in multiple places on the cat's face, on the chin below the mouth, in four or more parallel rows above the upper lip on each side of the cat's nose, in tufts above the eye and on each cheek. These special sensory hairs are capable of movement, extremely sensitive, and assist the cat to move accurately and maneuver successfully in the dark.

cats have specially adapted cells in their nose that enable them to smell some things that humans cannot smell. Domestic cats also have a scent organ, the *vomeronasal* or *Jacobson's organ*, which increases their ability to smell. To allow for air to pass the vomeronasal, a small cat must open its mouth and inhale. An open mouth with a dangling tongue exposed is referred to as *gaping*. Gaping is similar to *flehmen*, which is observed in big cats and wild sheep, especially during mating season.

What is special about a cat's sense of taste?

In 2005, scientists discovered that all cats lack one of a pair of proteins required to sense sweetness. The missing protein was the result of a deletion, the loss of part of a chromosome or sequence of DNA, in a gene. The deletion in the gene prevented one of the proteins that enables the perception of sweetness from being expressed (created). Because all members of the Felidae share this trait—the missing protein—scientists believe that this trait occurred very early in the feline evolutionary tree. The lack of ability to taste sweetness may have led early members of the Felidae to ignore plants entirely and become only hypercarnivores—obligate meat eaters.

Usually the tongue is associated with the sense of taste. A cat's tongue however, has multiple functions that go well beyond taste. Anyone who has been licked by a domestic cat immediately feels the rasping, sandpaper-like surface of the cat's tongue. Like all cats, the ocelot's tongue shares these same sandpaper-like features. A cat's tongue has backward pointing papillae on the surface. Papillae are rigid because they contain keratin that is

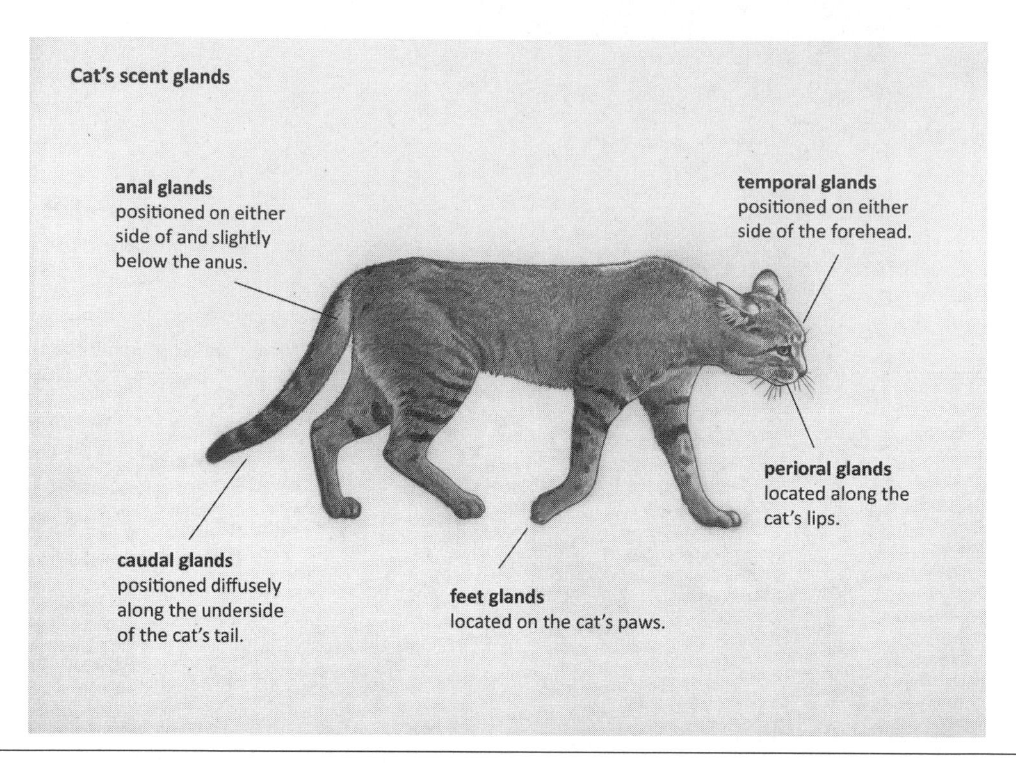

Cats communicate in various ways, but it is through scent marking that a stranger or potential mate will obtain messages about the resident male, or in turn passes on their own. Marking is in the form of an odorous substance which is passed with the cat's urine, feces, and saliva, or as illustrated, by special scent glands on the cat's body.

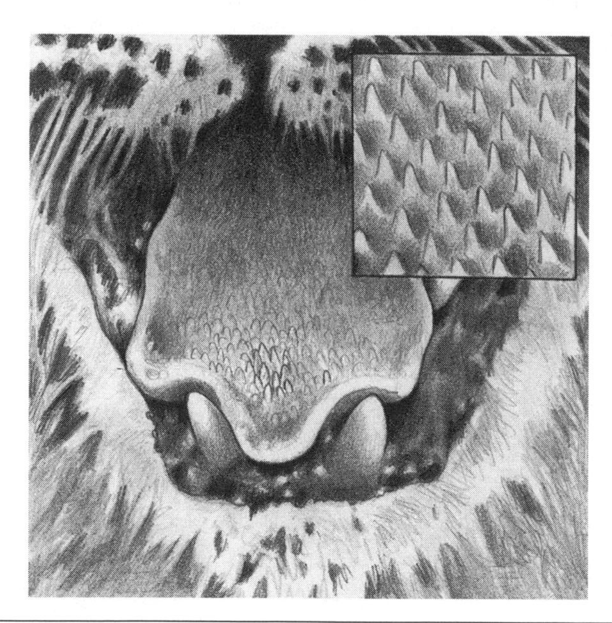

The cat's tongue is a multipurpose tool with a rough upper surface made up of abrasive "papillae" *(inset)*. **These backward-facing hooks serve to assist the cat remove hair, skin, and meat from its prey. The tongue is also an effective muscular ladle ideal for lapping up fluids or the perfect tool for cleaning the cat's body.**

also found in claws and human fingernails. Papillae aid in grooming and in rasping meat from bone.

Can you determine whether a fossil small cat was terrestrial or arboreal?

Arboreal small cats share at least two morphological traits: relatively large forepaws and long, relatively heavy tails for their body size. The clouded leopard and Sunda clouded leopard have stout limbs. The bones associated with stout and large limbs and tails are heavier and more robust, thus aiding a determination of whether or not a fossil small cat was terrestrial or arboreal. Unfortunately, the fossil record of the Felidae leaves much to be desired. As top predators, small cats are naturally rare and thus fossils are relatively rarer. Moreover, fossils that contain the necessary information such as the tail, forepaw, and foreleg bones are exceedingly rare. Typically, the relatively thicker skull and rock-hard teeth are fossilized and then later recovered, making determination of whether a fossil small cat was terrestrial or arboreal difficult to determine.

Chapter 3

Small Cat Colors

Can small cats be black?

A cat's fur color and pattern vary greatly, even among littermates. Domestic cats can be white, black, and many colors in between, even in the same litter. Small cats born in the same litter can be normally spotted and also all black. All black guignas are known to occur throughout their geographic range in Chile and Argentina. Indeed, when JGS studied guignas in Chile, one of every four guignas he captured and fitted with a radio-collar was black.

The most widespread small cat in North America is the bobcat. Although bobcats are rarely seen, occasionally we are lucky enough to get a fleeting glimpse of one. Bobcats, like many other small cats, can be all black. The black color of the fur is the result of an increased amount of melanin, a black pigment, in the fur. Small cats that are all black are referred to as *melanistic*. As far as is known all small cats can be black.

In addition to guignas, all black or melanistic Geoffroy's cats, pampas cats, tigrinas, Asiatic golden cats, marbled cats, and servals have been observed or photographed by hidden cameras.

What causes the different coat colors of small cats?

The fur of a small cat is the same as hair on humans. Hair is a thin strand of proteins that grows from hair follicles in the dermis of the skin and grows through the epidermis or outer layer of skin. During hair growth, pigment cells form melanin that determines color in each hair. As each hair grows, different colors can be deposited. Thus, each single hair can have different

Most small cats are known to have an all black, or melanistic, morph. This melanistic female guigna, photographed in Chile, has a lizard in her mouth.

colors. For instance, the hair of the Chinese mountain cat has a terminal tip that is light brown, a long mid-section that is dark brown, and a 2.5 cm (one inch) section of hair closest to the skin that is gray. In small cats with spots, like the ocelot, a hair can be all black and the adjacent hair can be orange.

Melanin is a collection of chemical compounds. Different amounts of each compound give rise to different colors. Two compounds, eumelanin and phaeomelanin, are found in melanin. The pigment eumelanin can be brown-black, and the pigment phaeomelanin found in hair gives rise to a reddish color. A mixture of the two creates different color combinations. Note that there are no perfectly camouflaged forest green small cats because the color green cannot be generated from a combination of eumelanin and phacomclanin.

Why does the length and thickness of small cat's coat vary?

The variation in length and thickness of a small cat's coat is a response to environmental conditions. Through selection, the biological process of evolution has produced small cats that are well-suited to their environment. The explanation of how this happens is quite simple. First, small cats (and all living organisms) do not adapt to their environment. Instead, those individuals that are best adapted produce more offspring. These offspring are

Table 3.1. Small cat color and markings

Species	Basic ground color	Coat markings
Asiatic golden cat	Golden brown to dark brown, red, or gray	Plain or marked with spots and stripes
bay cat	Bright chestnut	Faint spots on the legs and under parts
marbled cat	Dark brownish gray, yellow gray, or red-brown	Dark blotches, stripes, and spots
African golden cat	Orange to gray	Plain, partially, or completely spotted
caracal	Reddish brown	No markings
serval	Tawny gold	Black uneven stripes and spots
black-footed cat	Tawny gold	Black or dark brown spots that merge into bands/rings
domestic cat	Wide range of colors	None or variations of stripes, spots, and blotches
jungle cat	Sandy-brown, reddish, or gray	Faint stripes
sand cat	Pale sandy-brown to light brown	Indistinct dark bars
wildcat	Light ochre, yellow, or slate gray and dark mid-brown	Tabby markings of spots, stripes, and bars
fishing cat	Gray, olive brown	Small black spots
flat-headed cat	Rich rusty red to dark brown with silver tinge	No spots or stripes
leopard cat	Pale brown, bright reddish to gray	Spotted coat, dark spots, bands, or blotches
Pallas' cat	Grayish to reddish orange	Dark stripes
rusty-spotted cat	Brownish-gray with reddish tinge	Elongated brown blotches, large dark spots
bobcat	Light gray to reddish brown	Black or dark brown spots or bars
Canada lynx	Yellowish brown	Mottled coat of dark brown spots
Eurasian lynx	Fawn yellow gray to light brown	Mottled light reddish brown or dark brown spots
Iberian lynx	Pale yellow beige to light reddish brown	Heavy spotted, dark spots and bars
Andean cat	Silver gray	Dark brown spots and stripes
Geoffroy's cat	Brilliant ochre (north), silver gray (south)	Round black dots that merge into stripes
guigna	Buff or gray brown	Heavily marked with spots
margay	Yellow gray to tawny yellow buff	Dark brown or black spots, some with shaded centers
ocelot	Pale gray, reddish gray, or grayish yellow cinnamon	Solid and open dark spots and bars
pampas cat	Pale silver gray	Almost unmarked or red-gray spots and stripes
tigrina	Fawn or tawny buff	Black or dark brown spots and rosettes
clouded leopard	Grayish beige, light yellow ochre to yellow brown	Clouded shaped markings, black lines, and spots
Sunda clouded leopard	Grayish beige, light yellow ochre to yellow brown	Clouded shaped markings, black lines, and spots
jaguarundi	Red-brown or gray	No markings

more likely to be born well-adapted, because at least one parent was well-adapted to the local environment. Those that are well-adapted are more likely to survive (they are selected for) and produce more offspring, and those that are not well-adapted are more likely to perish (they are selected against) and thus produce fewer offspring. In this way, small cats evolve to be well-adapted to their local environment.

The process of evolution has produced the Andean cat, which is ideally suited for living in the high elevation, cold, and dry of the Andes of Argentina, Bolivia, Chile, and Peru. The Andean cat has very long, thick, and dense fur, and a long, bushy, feather light tail that serves as a scarf that covers the face during periods of rest.

The sand cat that inhabits the deserts of North Africa and the Middle East has short, sandy-colored fur—the color of sand—and fur that grows over the pads of the paws and toes, presumably to protect the paws and toes from the hot desert sand.

How are hair colors determined genetically?

The production and coloring of each hair is a complex, interesting, and active area of research. New results continue to increase our understanding of how hair is produced and colored. The more that is learned, the more scientists realize there is yet more to learn. Genetic determination of hair color in small cats has been investigated by Stephen O'Brien and Eduardo Eizirik at the National Cancer Institute in Fredericksburg, Maryland. However, most of what is known regarding genetic control of hair color comes from the study of laboratory mice. Scientists assume that hair production is similar in all mammals. If such an assumption is true, then it makes more sense to study genetic control of hair color in laboratory mice than in small cats.

Several genes control the deposition of melanin in each individual hair. Control is exercised in every possible way. It's not hard to imagine how a human artist would paint each individual hair follicle so that the collection of hairs would resemble a small cat's fur. Each hair is like an artist's easel; different combinations of melanin are the artist's paint. The artist in this case is a collection of genes. The genes control the movement of melanin-producing cells in each hair follicle, the location where the melanin is deposited, the combination of melanin that results in the color of the hair, the shape of the melanin, and whether the melanin cells are dispersed or clumped in each individual hair. All the hairs together give fur its look and feel, its texture, and its colors.

Just as artists have names, genes also have names. One of the hair color genes is named agouti (*A*). Like all genes, agouti (*A*) is made up of alleles.

There are several different versions (or alleles) of agouta (*A*). For instance, agouti allele (*A*), the normal agouti (*A*), produces banded hairs. The non-agouti allele (*a*) causes all hair and hence the coat to be black. From an artist's perspective, it's not hard to see how a hair might be banded. Because the hair grows continuously, the artist simply paints part of the hair with one color and then changes the color. By repeatedly alternating the mixture of paint, the melanin, a banded hair is produced. Genes serving the role of an artist emulate this same procedure.

A genetic defect is known to cause no melanin to be produced by the body. If no melanin is produced, then no color is produced and the hair and eye pupils appear white. In fact, the color we know as white is the wavelength of light received by our eyes that is reflected from hair that has no melanin. Albinism is known in many mammalian species but is exceedingly rare in natural populations. Native Americans of the Great Plains held that white bison, so-called white buffalo, were sacred. Albinism is known to cause poor eyesight, and albino animals are very conspicuous. Poor eyesight and lack of concealment are two fatal flaws for a small cat that uses its eyes to locate prey and stealth to stalk its prey.

Although albino, or white, tigers are known, most have been born in captivity. Contrary to popular legend, white tigers were not supremely adapted to Ice Age conditions. However, one very large predator is indeed white and is ideally camouflaged for its habitat—the polar bear. Polar bears are naturally white and albinism is rare, probably because polar bears with poor eyesight do not live long enough to reproduce.

Is there a reason for the patterns on the coat?

Because of the variety of coat colors and patterns, it might appear that coat patterns and colors are random. Indeed, spotted small cats don't have mirror images of the same spots on either side of their bodies. If in fact there is no reason why a coat takes on the color and patterns that it assumes, then it must be random. There is some truth to this belief.

As with humans, each individual small cat is made up of two sets of genes: one derived from the father and one derived from the mother. At this level, whether the gene comes from the father or the mother is indeed random. Once the genes are assembled, the individual's genotype is established and hence the individual's phenotype (the observable manifestation, or morphology, of the genotype), including its coat patterns, is determined. Because an individual's coat pattern is determined genetically, the coat patterns have been and continue to be subjected to evolutionary selection pressures. These selection pressures ultimately led to the coat patterns on small cats we see today. We know that similar but not exactly the same

Many small cats show different color morphs. These African golden cats show a golden morph and a tawny morph. Photo © Neville Buck

selection pressures produce similar but not exactly the same coat patterns on small cats, even on different continents.

For example, the ocelot of the Americas and the leopard cat of Southeast Asia both live in forested environments. Both are principally nocturnal—active at night. Both the ocelot and the leopard cat prey upon a variety of species, and rodents are a popular prey item. Because the ocelot and leopard cat occur on different continents, the prey species they consume are different species. Although not identical, the habitats these small cats occupy are similar, and the environmental pressures they face are similar. Because the selection pressures faced by the ocelot and leopard cat are similar, evolution has produced similar results: the ocelot and the leopard cat have similar spot patterns.

The spot patterns of the coat worn by each individual ocelot and leopard cat are, however, unique to the individual and, in fact, unique to each side of the individual. Thus, evolution has produced similar answers to similar problems. At the scale of the individual, the coat pattern is the result of a purely random process and so, within the bounds established by an individual's spotted parents, is unique.

Other examples serve to illustrate how evolution works. The Andean cat that lives in the high Andes and Pallas' cat that occupies the vast Asian steppe both live in rocky, cold, wind-swept environments. Both prey primarily upon small mammals. Selection pressures are similar in these habitats.

Both the Andean cat and Pallas' cat have thick, dense, long slate-gray coats and large bushy tails. Independent selection pressures produced similar but not identical results.

The pampas cat of South America shows that when selection pressures differ for a single species occupying different habitats, coat patterns differ. The pampas cat occurs in the great grasslands, the cerrado, of Brazil, and the Andes from Ecuador, Peru, and Bolivia to southern Chile and Argentina. Environmental selection pressures vary greatly from the low altitude grasslands of Brazil to the harsh, high and dry, wind-swept Andes. In Brazil, the pampas cat has a mostly brown coat without spots or stripes; in the Andes the coat of the pampas cat is beautifully spotted. Widely different environmental pressures in different habitats have produced different coat patterns—*in the same species*. A cat is a cat, but knowing precisely the particular species to which the cat belongs often requires an authority.

Are there age-related differences in coat color?

All mammals exhibit important and obvious morphological age-related differences. For instance, as humans age their hair color lightens. The backs of male gorillas begin to turn white around age twelve. Only then do we refer to them as silver-backs. Coat colors in some small cats show age-related differences. Young pumas are born with large spots that slowly disappear with age. Presumably spots aid in camouflaging the young kittens. In other cases, spot patterns remain the same throughout a cat's life. This means that a wonderfully spotted margay is born with the same number, placement, and color of the spots and stripes that it will keep throughout its life. The spots and stripes simply enlarge, or scale, as the young kittens age.

Scientists believe that spots provide camouflage. Camouflage is useful for hiding, both in defense and in stalking prey. Since each small cat's coat pattern is unique, camouflage may also aid identification. When a litter of small cats is born, the mother must provide for them until the kittens are old enough to hunt on their own. The mother must nourish not only her kittens, but herself as well. This means that she must leave the safety of the den to pursue prey. As the kittens age, they naturally become more inquisitive and venture out of the den on their own, especially when the mother is absent. At this time of their lives, they are most vulnerable to other predators, because they have little or no knowledge of the world outside the den. From an evolutionary perspective, it makes sense that kittens would be born with camouflaged coats.

African golden cat *(Caracal aurata)*.
Photo © Neville Buck

Andean cat *(Leopardus jacobita)*, Chile. Photo © Jim Sanderson

Asiatic golden cat
***(Catopuma temminckii)*, Thailand.**
Photo © Jim Sanderson

Bay cat *(Catopuma badia)*,
Sarawak, Malaysian Borneo.
Photo © Jim Sanderson

Black-footed cat *(Felis nigripes)*, South Africa. Photo © Alex Sliwa

Bobcat *(Lynx rufus)*, Arizona. Photo © Jim Sanderson

Canada lynx *(Lynx canadensis)*, Canada. Photo © Colorado Division of Wildlife

Caracal *(Caracal caracal)*, captive, Australia. Photo © Patrick Watson

Clouded leopard *(Neofelis nebulosa)*, Nepal. Photo © Alex Sliwa

Eurasian lynx *(Lynx lynx)*, Switzerland. Photo © Alex Sliwa

Fishing cat *(Prionailurus viverrinus)*, Thailand. Photo © Nancy Vandermey

Flat-headed cat *(Prionailurus planiceps)*, Thailand. Photo © Jim Sanderson

Geoffroy's cat *(Leopardus geoffroyi)*, Argentina. Photo by Lynn Culver

Guigna *(Leopardus guigna)*, Chile. Photo © Jim Sanderson

Iberian lynx *(Lynx pardinus)*, Spain. Photo © Alex Sliwa

Jaguarundi *(Puma yagouarondi)*, gray morph, Brazil. Photo © Jim Sanderson

Are there seasonal changes in coat color?

Seasonal differences vary with distance from the equator. Most places on earth experience four seasons. Near the equator, seasons differ more in the amount of rainfall than in temperature changes that are typical in higher latitudes. Thus, we might rephrase this question as follows: Does coat color respond to changes in rainfall or temperature?

Because most small cats have eluded in-depth long-term studies, scientists do not know if their coat color changes with respect to changes in local environmental variables such as temperature and rainfall. However, the pampas cat of South America does show exceptionally different coat color differences within its geographic range. In the great grasslands of Brazil the pampas cat has a spotless brown coat, with black stripes across the forelegs and a wide black stripe across the nose. In the high Andes the pampas cat's coat is gray with rusty spots. However, the black stripes on the forelegs and across the nose are present. The leopard cat of Asia, which occurs on islands near the equator such as Sumatra, Java, Borneo, and the Philippines, has an orange-brown spotted coat and is smaller than its relatives on mainland Asia. In more northern and colder climates in Asia, the leopard cat's coat becomes noticeably browner in color and body size increases. Larger body size helps the leopard cat to hold body heat.

As for changes in temperature, as far is known and anticipated, coats of small cats are not expected to change color seasonally. If other mammals serve as examples, then we should indeed expect that the coats of summer are shed and replaced with more suitable, thicker, winter coats of the same color and pattern. But as the altitude increases and cold conditions prevail throughout the year, the coat is unlikely to become thinner in summer. Though it is expected that coats of small cats do not change color seasonally, those small cats living in higher latitudes and lower altitudes likely replace their summer coats and thicker winter coats of the same color and pattern.

Is there much geographic variation in small cat species?

Geographic variation refers to the differences between spatially segregated populations of a species. Within-species variation is found in many species of small cats. The variation we see in the morphology of small cats such as its weight, size, and coat color and pattern ultimately has a genetic basis. The evolutionary forces that create morphological variation are better understood today than in previous generations.

The evolutionary biologist Ernst Mayr wrote that variation occurs not only within populations of species but also between populations of the same species. Studies of fossils also show that there is geographic variation in time in the same general location. Geographic variation is now accepted to be a universal phenomenon in living organisms. For instance, even humans show great geographic variation in their physical, or phenotypic, traits. Just as physical characters differ geographically, so to do behavioral and ecological characters.

Phenotypic differences result from both genetic and environmental factors. Selection over time determines which phenotypic characters are preserved in a population and which disappear. The genotype gives rise to the morphology we observe. Each local population is the product of continuous selection that adjusts and adapts the population to its local environment. Differing local environments give rise to geographic variation, as each individual within a population is selected to prosper in its local environment. In this way, better-adapted individuals produce more offspring than those individuals that are less well adapted. The genes responsible for the locally best-suited phenotype are thus passed on the next generation.

Within a local population there is an average genetic composition, an average genotype, and variation about the average, because reproduction involves a random combination of genes. Each individual has a random composition, and this gives rise to variation within the population. Selection pressures work on this variation. Since the local environment is constantly changing, the mean genotype also changes.

Morphological variation within the small cats is well known and used to name subspecies. The pampas cat inhabiting the Brazilian cerrado has a different coat pattern than the same species living in the high Andes. Geoffroy's cats living in Bolivia weigh half as much as those living further south in the pampas of Argentina. By examining the DNA of Andean cats and pampas cats gleaned from feces collected in caves and rocky shelters in the Andes, Daniel Cossios of Peru can not only determine from which small cat species the feces came, but the general locality where the feces was found. Daniel's analysis shows clearly that the genotype of Andean cat populations and pampas cat populations varies from location to location in the Andes.

In guignas living in Chile there is variation in body weight between island and mainland populations, and probably from north to south in different populations.

Previously thought to be a single species, the clouded leopard of the Asian mainland and the Sunda clouded leopard found in Sumatra and Borneo are now recognized to be distinct species. Until the molecular work of Valerie Buckley-Beason, the morphologies of the two clouded leopards kept secret the remarkable fact that the genotypes of the two species showed greater differences than a tiger does from a lion.

The leopard cat shows great morphological differences across the varied habitats it occupies in Southeast Asia. The leopard cat of Iriomote

Of all the small Asian cats the leopard cat has the broadest geographic range. The eastern Russian, or Amur leopard cat, was once thought to be a distinct species because of its different coat pattern.

Island in Japan is dark gray while the leopard cat of the Philippines is orange with black ocelotated spots. In both places, the leopard cat is a nocturnal rodent specialist.

The understanding of geographic variation within and between species is fundamental to the conservation of small cat species. Conserving geographic variation within a species is essential to conserving not just the species itself but the evolutionary processes that create new species. Geographic variation clearly demonstrates that biological diversity of small cats cannot be preserved and maintained in zoos. Thus, a holistic approach that integrates local ecology and genetics is necessary, but not sufficient, to conserve species. Both the history of the population, as revealed by genetic analysis, and the adaptive differences, as revealed by natural history and other ecological factors, must be factored together to address conservation concerns.

Populations of small cats, not species as a whole, are the units of evolution and hence the units of conservation. Armed with the latest technological tools, the study of geographic variation within small cat species is a newly opened frontier where science and conservation efforts are long overdue.

Chapter 4

Small Cat Behavior

Are small cats social?

No. As far as is known, the social systems of all 30 small wild cat species are completely asocial. If those few small cat species that have been studied in the wild serve as a guide, all adult individuals of every small cat species likely live a solitary existence except during the short mating season. Outside the breeding season, interactions between same sex individuals of the same species, when they occur, are brief, full of sound-and-fury and sometimes blood, and so are best avoided.

In some species, the land tenure system has been well-studied. Our best understanding is that an adult male maintains an exclusive territory referred to as a *home range*. Within the male's home range live one or more adult females, whose home ranges are also exclusive and are necessarily smaller than the male's home range.

Within the male's home range, female's home ranges abut one another. Similarly, male's home ranges often abut one another. An individual's home range is actively defended and, if invaded by another individual of the same sex, violently defended. That is, males defend their home range against other males that abut their home range. Within the male's home range, adult females do the same. A male does not protect or feed females living within his home range. Adult males allow adult females to live within their home ranges simply for the purposes of breeding with them.

After rearing the young and teaching them to hunt, an adult female drives them from their natal area. This so-called *dispersal period* is the most dangerous time of a wild cat's life. Individuals are forced to navigate a dangerous landscape to establish their own home ranges in unoccupied, and hence, in all likelihood, less-than-optimal territory.

Sub-adult male small cats must move swiftly through any resident adult male's home range or risk death from an encounter. The last thing an adult male will permit is the establishment of a competitor within his home range. Sub-adult females will try to become established anywhere they can carve out an area sufficient enough to feed themselves. This might be within an adult male's home range. Although a resident adult female will defend her home range against encroachment, if there is sufficient space, the sub-adult can become established. The resident male takes no interest in such intra-female interactions. The females within his home range are for breeding purposes only.

The social system of the world's small cats is in direct contrast to other carnivores such as wild dogs, the family Canidae, some of whose species are highly social.

Do small cats fight?

As a last resort, small cats fight to defend their home range against encroachment. However, before physical contact is made, a variety of methods are used to avoid tooth-and-claw engagement. Small cats use urine, feces, and anal sac secretions to delineate, or *mark*, their home range boundary. These markings, more scent than sight, serve to warn their neighbors not to encroach. Small cats sometimes scream loudly at their neighbors not to encroach. If scent, sight, and sound do not discourage encroachment, direct engagement cannot be avoided. However, altercations, while fierce, do not often end in the severe injury or the death of competitors.

Like most highly territorial carnivores, small cats are well-adapted to inter-species fighting. The skin of a small cat is not tightly attached to the muscle below the skin. In fact, the cat's body is very loosely enclosed within the skin. If you've ever tried to get a grip on your house cat, you fully appreciate what is required. The muscles move and slip inside the skin. The cat seems to be able to rotate its body within its skin, and invariably squirms free. Moreover, the skin is covered with hair that seems to slide in your grasp. Similarly, though an enemy's teeth and claws might penetrate hair and skin, they will most likely not penetrate muscle. In this way, what might become a fatal grip is thwarted. Though fur might fly, and loud war screams might fill the air, the chances are both combatants will emerge without serious injury, retreat to their respective home ranges, and resume their constant border patrol.

How smart are small cats?

Animal intelligence has for many decades fascinated scientists in different fields of study. However, just how smart animals are remains a subject of debate. After all, were we to compare the physical prowess of a lion with a human in a one-on-one contest, no thinking person would bet on the human.

Small cat kittens enjoy playing just as much as domestic cat kittens do. Here Ivan, a captive Eurasian lynx, takes a flying leap at his toy. Staying alive is a serious business that leaves little time for adults to engage in play activity. Photo © Wendy Debbas

The small cats we see today are the product of more than 30 million years of evolution. This is far longer than humans (*Homo sapiens*) have existed on earth. Small cats have outlasted the entire lineage of saber-toothed cats, the last of whose species went extinct 12,000 years ago. Small cats have outlasted two other lineages of cats that have all gone extinct as well.

Intelligence in small cats must be considered from an evolutionary perspective. Selection ensures that those individuals best adapted to their local environment produce the most offspring. Over a long time period, the population of small cats has thus become well-adapted to the local environment they inhabit. The results produced by evolution are good enough for the local conditions. If local conditions change, so do the small cats inhabiting them.

There is no march of "progress" to increasingly more intelligent small cats. Instead, today's small cats are adapted as best they can be to current environmental conditions. If, in the near future, greater intelligence is required (whatever that is with respect to small cats), then individuals in the population would face different selection pressures that enable those with greater intelligence, whatever that might be, to produce more offspring, and those with less intelligence would die earlier and so produce fewer offspring. Over time, the populations would become "more intelligent."

Note that if conditions changed again and the newly evolved intelligence was no longer necessary, selection pressures for "greater intelligence" would diminish. The result would be a small cat population with less intelligence but better adapted to its local environment. This is true for all living organisms,

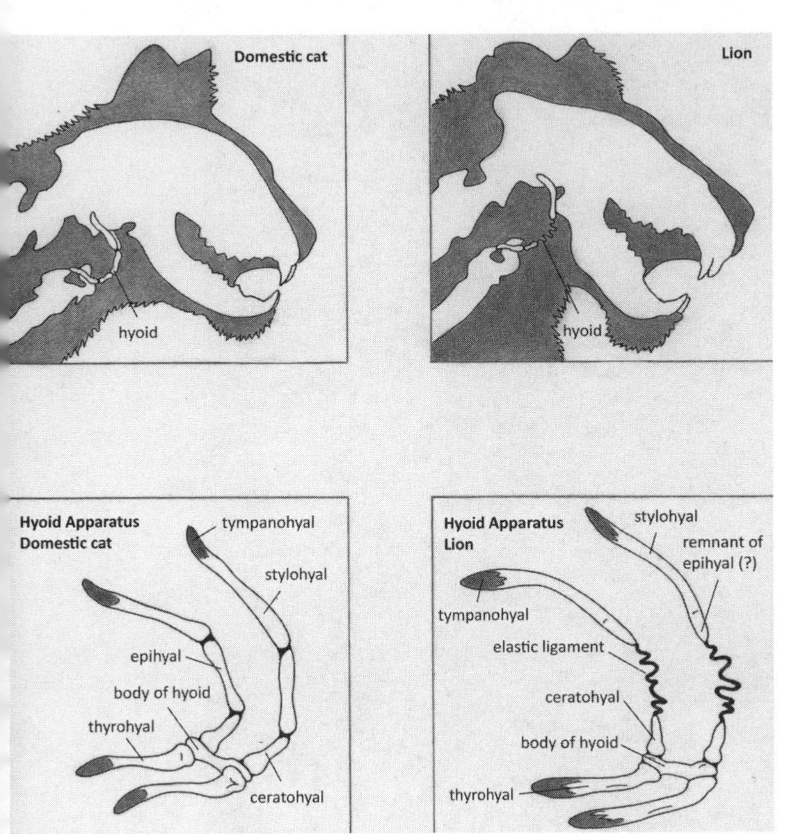

All cats make sounds using the hyoid bones and larynx or voice box (*left*). The hyoid bones of big cats such as lions are flexible and allow lions to roar while in small cats the bones are more rigid (*right*).

except humans, where environmental selection pressures have long since diminished in importance and in most places on earth, have nearly vanished.

Do small cats play?

Small cats, like most mammals, engage in what we humans consider play behavior. Most of what we know about small cat play behavior comes from a few small cat species that have been kept in captivity and inferences of what is known of play behavior in domestic cats.

Young small cats engage in play behavior individually, with siblings, or with the female adult parent. Such play behavior likely helps to develop their motor skills. Adults sometimes "play" with their prey. Adult Geoffroy's cats have been observed tossing captured mice into the air, stalking the lifeless or wounded mouse, and pouncing upon it, only to toss it again. Zoo-held guignas in Chile were observed playing with their food.

Different researchers have proposed several reasons why young small cats play. Most believe that play, such as handling objects like twigs or leaves, mouthing or biting objects, chewing on objects, rolling on their backs and using all four paws to claw at objects develops and reinforces behaviors that will be necessary when the young become adults.

Threatened by a domestic dog, a frightened guigna in Chile takes refuge in the nearest tree and closes its eyes so as not to attract attention.

Do small cats talk?

Small cats communicate in several ways but not in the same way or amount that humans do. Humans are the only species that use true speech to communicate. This is not to suggest, however, that small cats do not vocalize. They indeed do!

How do small cats avoid predators?

Small cats have several predators, both terrestrial and avian, and are skilled in predator avoidance techniques. However, because most small cat species have never been studied in the wild, not much is known of each species specifically. Most of our understanding is inferred from a handful of captive species, and thus much remains to be learned.

Small cats avoid predators by being constantly aware of their surroundings, cautious in their actions, and skilled at instant retaliation. Most small cats are able to climb trees with ease and often do so when threatened.

Chapter 5

Small Cat Ecology

Where do small cats sleep?

All small cats seek shelter to rest and sleep. This is in contrast to lions that sleep under open sky on the African savanna. However, lions are social and so sleep in groups that act as a form of protection. All small cats are solitary, and most have natural predators, so it's important for them to rest or sleep in a protected or easily defended place. Small cats don't dig so do not excavate their own burrows. Instead, they let another animal do the digging and then move in if the burrow is unoccupied, or they can evict the resident.

Often it's possible to look at a landscape and determine where sheltered resting places are found. For instance, in the Andes of South America there is very little vegetation, nights are extremely cold, and winds can be vicious. It's not surprising then that the Andean cat seeks out well-protected caves with sandy bottoms. Often these caves have many tens or hundreds of Andean cat feces. The Andean cat apparently uses its den as a latrine. Why would the Andean cat do this?

Vicuñas (*Vicugna vicugna*) are camelids living in the Andes. Vicuñas travel in small herds. It is well-known that vicuñas use communal latrines. The latrines are mounds of feces that individual members of the herd add to each day. At night, the vicuñas sleep on the latrine mound. The reason for this is obvious: it's warmer on the mound than elsewhere. The decaying feces produce warmth and the herd sleeps on this natural heater.

Protected from the wild by the cave walls, the Andean cat might also be taking advantage of the warmth of its decaying feces. With its thick, rich tail as a scarf, and crouched on a bed of sand to protect its paws from the cold, the Andean cat spends the night in its protected shelter.

It's not hard to imagine that a Pallas' cat, or manul, also sleeps in a cave shelter or a burrow in the rocky outcrops it inhabits on the Asian steppe. The Pallas' cat's tail might also function as a scarf.

The guigna, a forest cat found mostly in Chile, sleeps and rests in hollowed out fallen trees, under exposed stumps, in piles of sticks, inside dense bushes, and, where the cats shares the land with humans, in human-constructed shelters such as barns. They may even sleep under people's houses. Guignas easily climb trees and rest during the day on a tree branch, safe from feral dogs, their most dangerous predator.

The sand cat and the black-footed cat usually sleep during the hot day in a cool underground burrow excavated by another species. The Chinese mountain cat is a subspecies of the widespread wildcat. It inhabits the largely treeless Tibetan plateau, where badgers and marmots also live. Badgers and marmots are excellent excavators and live in deep burrows that, when abandoned, can be used by these cats.

Since many of the small cats have not been thoroughly studied in their native habitats, there are still many new facts to be discovered. Where do the flat-headed cat, bay cat, marbled cat, Asiatic golden cat, and African golden cat sleep? We can only make an educated guess based on what is available in the habitat and what we've learned from other small cats.

Do small cats migrate?

Small cats, like their big cousins, do not migrate seasonally. Migration typically refers to movements by a population or large group of individuals, not the movement of a single individual. Mammals that migrate are typically large, with either long legs or large flippers, and are responding to seasonal changes in their food supply. The food supply of small cats is very often rodents that are incapable of migrating, so small cats have no need to migrate.

Small cats also establish and defend territories until they die or are usurped, forcing them to be nomadic. The establishment of a more or less permanent defendable territory and lack of prey migration effectively makes wild cats homebodies. Though they often explore adjacent areas, they return to their territory quickly, where they know every nook and cranny where prey can be found, the best shelters, and where to hide when an intruder threatens.

The Eurasian lynx has been thoroughly studied in Switzerland by Urs Breitenmoser and his spouse, Christine Breitenmoser-Würsten. They have reported movements of over 200 km by individual Eurasian lynx.

Small cats like this Eurasian wildcat female and young often sleep in hollow logs and other secure places.

Photo © Neville Buck

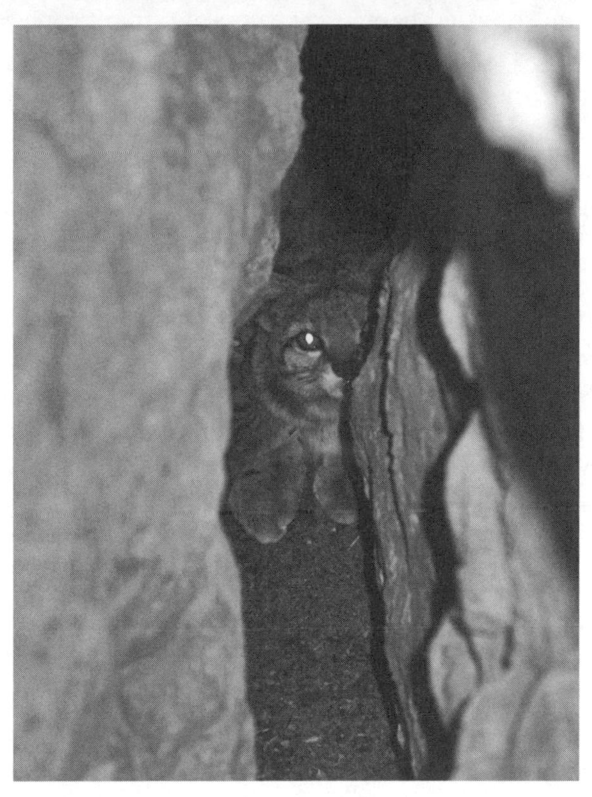

Secure in her cave den in the Bolivia Andes, a female Andean cat shows no fear of researchers.

However, despite the large distance, this was likely a dispersal event, a subadult male leaving his birthplace looking for favorable, unoccupied habitat. A dispersal event should not be considered a migration.

How many small cat species can coexist?

First, let's review what we know about small cats and their home ranges. Because small cats are solitary and are strongly territorial, same-sex individuals of a single species coexist by maintaining exclusive, non-overlapping home ranges. A male's home range is larger than a female's home range. Often multiple exclusive female home ranges are contained with a male's home range. Thus, the coexistence of individuals of a single species is at best in dynamic balance. Armed with this vital information, the question of coexistence of multiple species can be addressed.

Ecological theory holds that if two similar species compete for the same resources at the same time, one will prosper and one will disappear. Thus, coexistence depends on the resources available and the time those resources are exploited. A more varied habitat can support different species. For instance, a grassland savanna can best be utilized by a serval, and a forest might be occupied by an African golden cat. Both wild cats coexist, more or less, by occupying different habitats.

Suriname has six species of cats, four of which are small cats: jaguar, puma, ocelot, margay, tigrina, and jaguarundi. From 2006 to 2009, local colleagues and JGS ran remote cameras in a single forested area in Suriname. At many of the remote camera locations five species of cats—jaguar, puma, ocelot, margay, and jaguarundi—have been photographed. How do these five species coexist in a single area?

To avoid competing for the same resources, the ocelot, margay, and jaguarundi divide space and time between them. Some 98% of more than 300 independent photographs of ocelots show that ocelots are nocturnal—active at night—and terrestrial. More than 200 photographs of jaguarundi, many taken at the same locations that recorded ocelots, show they are diurnal—active during the day—and terrestrial. Margays are also primarily diurnal, however, margays are highly arboreal—they live in trees. These three small cats coexist in the same area and avoid competition by dividing time and occupying different habitats.

Small cat species coexist by exploiting similar resources at different times, or by occupying different habitats at the same time. This is summarized very simply: small cat species coexist by partitioning resources so as to avoid competition.

One way different small cat species coexist is by eating different food, being active at different times of the day or night, and by occupying different habitats. From top to bottom, the ocelot, Geoffroy's cat, margay, and tigrina are members of the genus *Leopardus*, whose members all lack a pair of chromosomes found in all other cats. Anne-Sophie Bertrand © 2007

Are small cats equally distributed throughout the world?

Small wild cats occur on five continents: Africa, Asia, Europe, North America, and South America. The domestic cat is found wherever humans live. In absolute terms without respect to land area, it's not surprising that the largest continent, Asia, has the greatest number of species and the smallest continent, Europe, has the least number. To see this, we must review which small cats occur on each of the five continents. (See also the small cat species distribution map in chapter 1.) Note that several species occur on more than one continent or geographic area, for example, the ocelot occurs in North and South America.

Africa has seven small cat species: African golden cat, black-footed cat, caracal, jungle cat, sand cat, serval, and wildcat. Asia has 14 species: Asiatic golden cat, bay cat, caracal, clouded leopard, Eurasian lynx, fishing cat, flat-headed cat, jungle cat, leopard cat, marbled cat, Pallas' cat, rusty-spotted cat, Sunda clouded leopard, and wildcat. Europe has just three species: Eurasian lynx, Iberian lynx, and wildcat. North America has six species: bobcat, Canada lynx, jaguarundi, margay, ocelot, and tigrina. South America has eight species: Andean cat, Geoffroy's cat, guigna, jaguarundi, margay, ocelot, pampas cat, and tigrina.

Because the same small cat species occur on adjacent continents, there appears to be many more small cats than exist. Asia has 14 small cat species; almost half the world's small cats and more than half of the world's cats occur in Asia, the largest continent. Africa has six species of small cats. North American and South America have six and eight small cat species respectively. Of North America's six small cat species, four occur in Central America. The smallest continent, Europe has just three species. From this one might surmise that the center of origin of small cats is likely Asia.

How do small cats survive in the desert?

Several small cats make a living in deserts. Bobcats occur in all four deserts of North America (Chihuahuan Desert, Mojave Desert, Sonoran Desert, and Great Basin Desert), the sand cat occurs across the Sahara Desert and the deserts of the Middle East, and the Pallas' cat is found in the arid Asian steppe. The Andean cat and pampas cat are also found in the Andes of South America that includes the Atacama Desert, where winters are very cold, summers are extremely hot, and years can pass without measurable rainfall.

Each of these small cats has special adaptations that enable them to survive. The Andean cat has a thick coat of long and rich fur, and a feather-light bushy tail that acts as a scarf. The Pallas' cat is similarly outfitted. The sand cat must endure extreme temperatures and a lack of water. The sand cat has long hair growing between its pads and toes that functions to protect the cat's paws from the excess heat of the desert sand. The bobcat is a remarkably adaptable cat that is found in Maine, where there are cold winters and deep snow, and in the Sonoran and Chihuahuan deserts of Mexico, where summer temperatures are very high.

How do small cats survive the winter?

Just as do house cats, all small cats shed their summer coat in the late fall. The light coat of summer is replaced with a rich, dense coat having a coarse outer layer and a dense fine layer that acts to trap body heat.

Table 5.1. Small cat habitat

Species	Habitat
Asiatic golden cat	Deciduous forests, tropical rainforests, and occasional open habitats
bay cat	Dense forests, areas of rocky limestone at the edge of the jungle
marbled cat	Forested areas, tropical, and rainforests
African golden cat	Moist forest, dense secondary vegetation along watercourses, drier open countryside, and alpine moorland
caracal	Dry woodland, savanna, acacia scrub, arid hilly steppe, and dry mountain
serval	Areas near water, open grasslands
black-footed cat	Dry open areas with rocks, scrubby bushes, and grass cover
domestic cat	Human habitation, or, in the case of feral cats, a place that affords adequate prey and a safe environment
jungle cat	Wet reed beds, arid scrub jungle, agricultural croplands, and dense forests
sand cat	Arid regions, rolling sand dunes, flat stony plains, and rocky deserts
wildcat	From deciduous and coniferous forests, open rocky ground, scrubby brush, and agricultural croplands
fishing cat	Areas of thick cover near water, marshes, mangrove, and dense vegetation
flat-headed cat	Along riverbanks in forested areas, also found in palm oil plantations
leopard cat	Dense tropical forests, pine forests, scrub, semi-desert, secondary vegetation, and agricultural areas
Pallas' cat	Desert, steppes, and treeless rocky mountainsides
rusty-spotted cat	Humid forests, low scrub, deciduous forests, scrub forests, grassland, and arid scrub
bobcat	Conifer and hardwood forests, brush, sage, and semi-desert areas
Canada lynx	Boreal forests, farmland, heavily wooded areas with adequate cover
Eurasian lynx	Forested areas with plenty of dense undergrowth and cover
Iberian lynx	Wooded areas, remote mountain regions, sand dunes, and scrub
Andean cat	Rocky treeless zones, high up in the Andes, cold, windy, and arid areas
Geoffroy's cat	Scrub woodlands, open bush, rocky terrain, and riverine forest
guigna	Coniferous forests, woodland areas, and semi-open habitats
margay	Heavily forested habitats, also found in coffee or cocoa plantations
ocelot	Tropical and subtropical forests, tropical evergreen, dry deciduous forests, dry scrub, and flooded savannas
pampas cat	Open grasslands, clouded forests, humid forests, prefers high altitude
tigrina	Cloud forests and humid lowland forests
clouded leopard	Dense forest, tropical and subtropical forest up to an altitude of 3,000 meters
Sunda clouded leopard	Dense tropical and subtropical forest up to an altitude of 2,000 meters
jaguarundi	Tropical rainforest, arid thorn forest, dense second growth forest, swampy grasslands

Note: Small wild cats occur naturally on all continents except Antarctica and Australia and live in a variety of habitat types and climatic conditions. Only domestic cats are found globally and reside with humans.

Cats living in very hot or cold climates tend to have fur-covered foot pads that protect them against the scorching desert sands or harsh winter snows. These pads, in the center and the tip of the cat's toe, are extremely sensitive and cushion the feet when the cat is running. They are also the cat's true sweat glands.

Pallas' cat puts on a thick coat for the frigid winter in Mongolia. In summer the fur is shed for a less dense coat. Photo © Bill Swanson, Cincinnati Zoo and Botanical Garden

Do small cats hibernate?

No. The states of torpor, typically associated with some hummingbirds, and hibernation, typically but incorrectly associated with bears, evolved to enable animals to conserve energy by lowering their basal metabolic rate (the body's minimum energy consumption necessary to stay alive) in response to a lack of resources. When resources become scare it makes sense to reduce the metabolic rate to conserve energy. Some hummingbirds enter torpor at night when temperatures drop and reemerge when the temperature rises. Some mammals enter a state of torpor for months at a time by greatly reducing their metabolic rate. This usually happens in the high latitude winter or at very high elevations in the tropics when the elevated metabolic rate needed to maintain body temperature can no longer be supported by food resources.

Small cats do not hibernate. When small cats are not hunting or patrolling their territory, they are resting in their dens. Here two captive sand cats rest during the day.

Most prey of small cats do not hibernate; therefore small cats do not enter torpor or hibernate because one or more species of prey are available to them year round.

Do small cats have enemies?

Yes, small cats face a variety of enemies. Small cats must constantly be on the lookout for a wide variety of aerial, terrestrial, and aquatic predators, depending upon where they live. Eagles, owls, and other large birds of prey will not hesitate to take small cat kittens. Big cats and feral dogs are known predators of small cats. Crocodilians (caiman, alligator, and crocodile) are able to prey upon small cats that venture unsuspectingly down to the river for a drink. However, small cats are extremely fast and nimble, and most are skilled, speedy climbers. When surprised by predators, small cats jump upward and backward in a fraction of a second. Contrary to legend, small cats have but one life and as a consequence are ever-alert for their enemies.

Do small cats commit infanticide?

Yes, small cats occasionally commit infanticide. Infanticide is the intentional killing of a young or newborn kitten. Infanticide has been documented in the big cats and in domestic cats, so it's natural to assume small cats also commit infanticide. Infanticide typically happens when the resident male that likely fathered the kittens is replaced by a new male. The killing of the kittens brings the female into a receptive condition that allows the new male to breed with her and hence father a new generation.

Do small cats get sick?

Like all mammals, small cats are susceptible to injuries and diseases that can make them ill and ultimately result in death. Scientists have suggested that a small cat's purring is more than just a soothing relaxing sound. A cat's purr happens at the precise frequency that aids the healing of broken bones, whose slow healing might result in a greatly weakened individual more susceptible to diseases.

In 2003, scientists at Fauna Communications Institute of Hillsborough, North Carolina, discovered that low decibel vibrations between 20 and 140 Hz (cycles per second) help heal bone fractures, repair torn muscles and ligaments, reduce swollen soft tissue, and relieve pain. Remarkably, cats purr at frequencies between 25 and 50 Hz—near optimal frequencies for bone growth and fracture healing. All cats, including large cats like pumas and lions, generate low-level sounds that increase muscle strength, joint mobility, and decrease pain. A Veterinary Association survey of cats that had plummeted from multiple floor buildings recorded that 90% survived including one that fell 45 floors. Veterinary doctors report that cats purr, even when traumatized or severely injured, and when giving birth.

No morphological characters, such as a cat's purr, that have been subjected to more than 30 million years of evolution, however small and irrelevant they seem, can be considered superfluous, especially those that we humans take for granted and do not fully understand.

How do small cats influence vegetation?

Small cats are predators, sometimes top predators, and so have an indirect impact on their habitat and hence vegetation. Predators have been repeatedly shown to maintain higher community biodiversity by preying upon those species that are most abundant. In the absence of predators, ecosystems have reduced biodiversity because they are dominated by the strongest competitors.

Small cats prey upon the most dominant and abundant species. Typically, prey species are herbivores or omnivores that consume vegetation. The loss of a small cat would thus produce a cascade of changes in the ecosystem. Because of predation by small cats, more species are able to carve out a living, even when one or a few prey species out-compete other prey species for limited resources. If small cats were removed from an area, over time there would be a quantifiable reduction of biodiversity. One change would be the loss of less competitive and probably rarer species, whose presence was maintained by a higher predation rate of a more dominant and numerous species. The resulting change in prey relative abundances and species losses would, over time, cause a change in vegetation.

In ecosystems where small cats occur, small cats maintain high biodiversity. If a small cat species is eliminated from an ecosystem, the ecosystem will experience a loss of biodiversity greater than the loss of the small cat. Changes in vegetation would take place as a result of changes in the herbivorous species the cats preyed upon.

Chapter 6

Reproduction and Development of Small Cats

How do small cats reproduce?

Most of what we know about reproduction in small cats comes from those small cats held in captivity. Mating is rarely observed in the wild, and for those species never held in captivity, everything is educated speculation.

All adult small cats live solitary lives, except during the brief mating season. In most species, the resident male is attracted to an estrous female by different cues, including her behavior, scent markings, and vocalizations. The male constantly checks the breeding condition of the females within his territory during this time. He also patrols his territorial border to prevent other males from breeding with the females within his territory. For males, the breeding season is brief but busy.

The length of time females remain in breeding condition, called *heat*, is not known. In domestic cats, females can be in heat every two weeks for nine months. However, in most small cat species, females are in heat at a specific time of the year and only for a few weeks.

The actual mating takes place when the male mounts the female. Typically the male bites the scruff of the female's neck. Since the skin of a cat is loose fitting, the male's biting does not do permanent damage. The male domestic cat's penis has more than 100 backward-pointing spines about 0.7 mm (1⁄5 in) long. Constanza Napolitano and her colleagues in Chile have discovered male guignas also has similar spines, and similar spines are anticipated to be identified in other small cat species.

Penis spines are suspected of causing ovulation in the female. Ovulation is believed to occur when the spines scratch and irritate the walls of

A male ocelot copulating with an adult female. Note that the male is biting the neck of the female. Such biting is typical and, because the skin of the cat fits loosely, such bites are not usually harmful. Photo © Neville Buck

the female's vagina. It is known that a domestic cat female may not become pregnant with each mating. Domestic cat females are superfecund: during a single heat females may mate with multiple males and can have different kittens from different males *in the same litter*.

Because small cats are highly territorial, males likely repeatedly mate with the females within their territory. This serves two purposes. First, the more time they spend in the female's presence, the less time other non-resident males have an opportunity to mate with the female. Second, his chances of siring all the offspring increases with each mating. However, because a resident male likely has multiple females living in his territory, the male is constantly busy. In his absence, the female will not hesitate to mate with other males that penetrate the resident male's territory. The resident male is also attempting to mate with nonresident females in another male's territory.

How long are female small cats pregnant?

The gestation period, the number of days a female is pregnant, for domestic cats ranges from 60 to 70 days. The gestation period for all other small cats is known only from those small cats kept in captivity. The small sand cat has a gestation period of from 59 to 63 days, the black-footed cat average about 68 days, and the much larger clouded leopard has a gestation period of from 86 to 92 days. All the small cats for which we have information range from around 59 days to 92 days; the average is approximately 75 days.

The male cat's reproductive system consists of the penis and a pair of testicles located outside the abdominal cavity, below the anus and enclosed within the scrotum. The testes produce the male hormone *testosterone*, as well as the sperm, which require a temperature slightly cooler than the cat's body. Sperm become fully developed as they pass down the tubes coiled over the surface of the testicles.

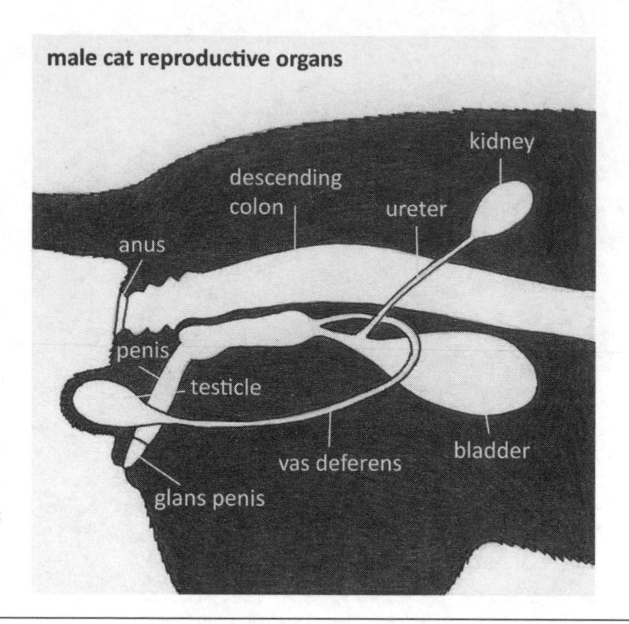

The female's reproductive system is positioned within the abdominal cavity with only the *vulva*, or outer labia of the female cat's genitals being visible. Basically the female reproductive system revolves around the pair of ovaries, which are positioned on either side of the cat's spine just behind the kidneys. The female hormone *estrogen* is produced by the ovaries following the stimulus of mating; *ova* or eggs maybe expelled at this time.

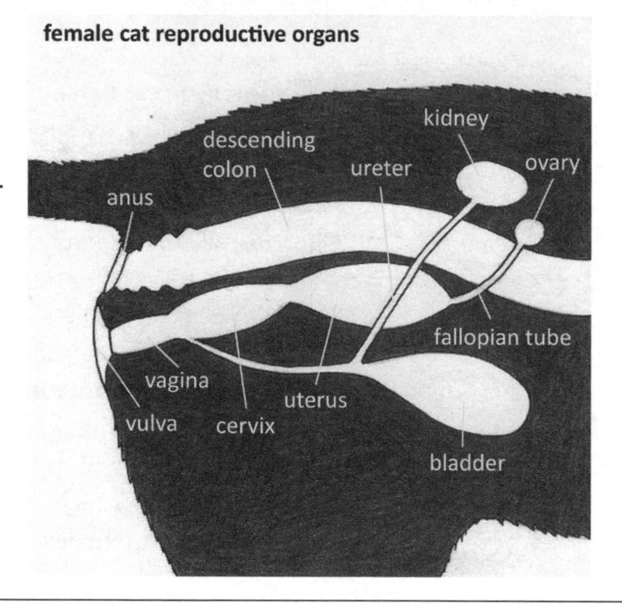

The gestation period is proportional to the body weight—the smallest small cats have the shortest gestation periods and the largest small cats have the longest gestation periods. See table 6.1 for gestation periods and the typical number of kittens born in a litter for those small cats for which we have information.

Where do mother small cats give birth?

Small cats are naturally secretive, and female small cats about to give birth are exceptionally shy and wary. From those small cats kept in captivity

Table 6.1. Small cat gestation, number of young, and den site

Species	Gestation period (in days)	Number of young	Den site
Asiatic golden cat	75–80	1–2 (3)	?
bay cat	?	?	?
marbled cat	66–82*	2*	?
African golden cat	75*	2*	?
caracal	68–81	1–3 (6)	Caves, tree cavities, or burrows dug by other animals
serval	67–77	1–2 (4)	Disused burrows or dense vegetation
black-footed cat	63–68	1–4	Disused burrows and termite mounds
domestic cat	60–70	3–6 (10)	Thickets, rock piles, or any well-hidden dry place•
jungle cat	63–66	2–4 (6)	Burrows, roots of trees, and hollow logs
sand cat	59–66	2–4	Abandoned burrows and thought to dig its own
wildcat	63–70	2–5 (8)	Underground den, rock crevices, and under bushes
fishing cat	63–70	2–4	Rough nest in dense patches of reeds
flat-headed cat	?	?	?
leopard cat	63–72	2–4	?
Pallas' cat	66–75*	1–4*	Rock fissures, rock crevices, and abandoned burrows
rusty-spotted cat	66–70*	1–2*	?
bobcat	60–70	2–4 (6)	Hollow logs, small caves, and rock crevices
Canada lynx	63–70	2–4	Within hollow logs of fallen trees
Eurasian lynx	63–74	1–4	Roots of trees, rock piles, and crevices
Iberian lynx	63–73	1–4	Small caves, under rock ledges, hollow logs, and fallen trees
Andean cat	?	?	?
Geoffroy's cat	62–67*	1–3*	?
guigna	?	?	?
margay	76–84	1–2	?
ocelot	79–82	1–3 (4)	Dense thickets and grass tussocks
pampas cat	70*	1 – 3*	?
tigrina	75–76*	1–2*	?
clouded leopard	85–95*	1–5 (2)*	?
Sunda clouded leopard	85–95*	1–5 (2)*	?
jaguarundi	60–75	1–4	Hollow trees, dense thickets, and overgrown ditches

Note: Gestation period and typical number of young vary between species. For most mammals, gestation period is proportional to the average body weight. Thus, heavier small cats have longer gestation periods than smaller small cats. The average number of young born is approximately one-half the number of nipples a species has.

Figures in parentheses indicate maximum numbers. *Figures shown from captive cats. ? = No data available.

•Indicates data for feral cat

Small cats give birth in secure, hidden dens. Here two very small caracal kittens rest inside a fallen log.
Photo © Neville Buck

Caracal mother carries her kitten in her mouth, a typical feline method for transporting the young. The mother will often move her kittens in this way when she changes the den location. Photo © Neville Buck

and the few that have been observed in the wild, we know they give birth in well-sheltered hidden dens located in hollow logs, underground burrows, under rocks, in caves, and even in abandoned human-made structures.

In 2004, JGS and his colleagues observed an Andean cat female and kitten in a well-hidden cave on the top of a cliff face in the Bolivian Andes. Held securely by his colleagues, JGS was able to hang over the edge of the precipice to view into the dark cave. The floor of the cave was sand. Though it was about noon and the sun was high and bright in the sky, he had to use a flashlight to peer into the cave. Two sets of green eyes stared back at him—an adult female Andean cat and her kitten. This was one of the most incredible experiences he had ever had—to be no more than 3 m (9 ft) away from so rare and beautiful small cats.

How many babies do small cats have?

The average litter size of domestic cats is four kittens, but the number of kittens is highly variable. Table 6.1 shows the number of kittens that various small cat species typically have. Note that almost all that we know comes from those small cat species that are held in captivity. Observing a female small cat with kittens in the wild is an extremely unusual event. The average number for all small cat species is three kittens.

Are all littermates equally related?

No. Small cats are superfecund. This means that females can be successfully impregnated by more than one male during a single heat. Thus, though all the kittens are born together, kittens can be half-related if the female bred with more than one male.

Recall that in the wild a typical male's home range includes several different females. When a female enters heat, the resident male typically stays with her and breeds with her often. However, several of the resident females might enter heat simultaneously. Such an event makes it necessary for the male to visit each female as frequently as he can while at the same time defending incursions by other neighboring males. Since his home range boundary is constantly being probed by neighboring males, it's not unexpected that while the resident male is attending one female, another nonresident male is breeding with another female.

Of course, if the female did indeed mate successfully with different males, there is no way of knowing which male sired which kittens. Though the kittens are indeed fully related to the female, it could well be that only some of them are related to the resident male. Moreover, there is no way for the resident male to know which kittens, if any, were sired by him.

A female Iberian lynx nurtures her kitten. The Iberian lynx is Critically Endangered and fewer than two hundred remain in the wild. A successful captive breeding program in Spain will help supplement wild populations. Photo © Alex Sliwa

How long do female small cats nurse their young?

Domestic cats nurse their young for up to seven weeks. Other small cats can nurse their young for two to six months. Small cats typically begin eating solid food at four weeks. Because most of what we know comes from those small cats held in captivity, and because captive situations are so much different from conditions in the wild, much of what we know could in fact be a very rough approximation to fact. Even when studies of small cats have been done in the wild, few researchers have observed females with kittens, and fewer still have documented mating, gestation, the birth of the kittens, and weaning. Most of what we know comes from small cats held in captivity and informed speculation.

How fast do small cats grow?

Domestic cat females can reach sexual maturity at five to ten months; males mature more rapidly and can reach sexual maturity in five to seven months. On average, most small cats are 80% full grown within a year.

A female caracal nurses her kittens. By the time the kittens are eight weeks old, they will be eating solid food. Photo © Neville Buck

Staying close to its mother, this young Pallas' cat, like all young small cats, is learning the important lessons of life. In the wild its very survival depends on how well it observes and mimic's its mother's actions. Photo © Neville Buck

They will add a little more muscle mass the next year, but their skeletal frame is pretty much finished growing by the first year.

How long do small cats live?

Domestic cats can live 15 to 22 years, and even more. Elderly domestic cats have been discovered by their owners to be totally blind but are able to go about the house without the slightest difficulty. However, life in a house is a luxury reserved only for domestic cats. Life in the wild is, as Thomas Hobbs wrote in *Leviathan* (1651), "nasty, brutish, and short," though he was not referring specifically to small cats.

In nature, small cats face many dangers and, more often than not, most die an early death. From incompetent mothering, to careless adventuring, and the vagaries of the weather and food supply, kittens are incredibly vulnerable. If they can avoid becoming food for another animal or bird of prey, they become sub-adults—big enough to get into more serious trouble.

Once small cats reach maturity, they are forced by their mother to secure food on their own. Unable to take up residence within the parent's territory, they must disperse to another area. Dispersal is fraught with multiple dangers. The dispersing individuals must go it alone into unknown areas they have never before explored. If the area has established adults, the adults will drive off the intruder, or kill them if they can.

Dispersers must find food in places they have never been, they must be elusive, and they must go quickly. Urs and Christine Breitenmoser radio-tracked a dispersing male Eurasian lynx in Switzerland that traveled over 160 km (100 miles) before taking up residence in an unoccupied territory.

Once in an established territory, the life of a small cat settles down to acquiring food, seeking mates, and defending its territory against those competitors seeking to usurp it. Life does not get any easier, but the skills necessary to live it have been tested and perfected. Nevertheless, the chances are that the individual will not die of old age. Rather, it will eventually be driven from its territory by a stronger competitor, at which time it will become a loner that ranges over a larger, more diffuse area, and avoids resident males.

While it is possible for small cat to reach 10 to 12 years of age, it is highly unlikely. Most will perish at half their maximum possible life span. If a small cat that no longer has a territory of its own can avoid being killed by a stronger male, it can survive longer. However, old age is not pretty. Over the course of a lifetime, tooth wear takes its toll, and the killing teeth, the canines, can break or wear down, making it difficult to secure food. Though competitors might be avoided, disease and starvation cannot.

Chapter 7

Foods and Feeding of Small Cats

What do small cats eat?

Many small cats are generalist predators that prey upon whatever they can catch. Other small cats are specialist predators whose diet mostly consists of a single prey species. The diet of a few small cats remains largely unknown. Knowing what a small cat eats is vital to conservation efforts. In the absence of this fundamental knowledge, the best planned conservation efforts may well end in failure. Preventing the extinction of a small cat species might be achieved by preventing the decline of its primary prey as is discussed below.

Usually a small cat's prey is smaller than they are. However, there are interesting exceptions to this general rule, and many small cats are quite capable of bringing down larger prey. Caracals are known to jump a 3 m (9 ft) fence and prey upon domestic sheep. Others, such as the Eurasian lynx, prey upon roe deer. Bobcats are capable of bringing down sub-adult deer and feral pigs.

These observations prompted some wild cat specialists to suggest that those wild cats that prey upon hoofed animals should be considered big cats. Those small cats that do not prey upon hoofed animals would then be considered small cats. However, much depends on the size of the hoofed animals and the relative sizes of the small cat and its prey. In Chile, several scientists and lay people have suggested that the diminutive guigna might occasionally prey upon young and sub-adult Chilean pudú (*Pudu puda*). Pudú have hoofs and are the world's smallest member of the deer family. Like other members of the deer family, Cervidae, adult male pudú have antlers.

The body weight of an adult male guigna is about 2.5 kg (5.5 lbs), about half the size of a domestic cat. An adult pudú stands about 38 cm (15 inches) at the shoulder, and weighs about 19–15 kg (20–33 lbs). Sub-adult and younger pudú are smaller and weigh less, but can easily out-weigh a guigna. If, in fact, guigna prey upon pudú, then, by the above definition suggested by a wild cat specialist, a guigna would be a big cat!

Many small cats such as the guigna are generalist predators that eat whatever is easiest for them to catch. For instance, the guigna preys upon insects, rodents, birds, lizards, frogs, an arboreal small marsupial known locally as monita del monte, and, much to the guigna's detriment, domestic chickens and geese. Like all generalist small cats, guignas hunt any time of the day or night. Anything they can catch is on their menu, and they show no preference for any item in particular.

Margays are presumed to be arboreal specialists that spend most of their time hunting in trees. Margays are probably generalist predators of bats, birds, lizards, and arboreal rodents; however, this is just conjecture because the diet of margays has not been analyzed. JGS and his Brazilian colleague, Alcides Renaldi, observed a radio-collared margay hunting by day in a tree. They wondered how so conspicuous a predator such as a margay might catch prey in trees during the day. The answer, once known, was obvious. They were surprised to observe the margay hunting roosting bats! The marbled cat's diet is not known, however, it too is highly arboreal and so might be similar in its feeding habits to the margay.

Other small cats such as the leopard cat, Andean cat, fishing cat, serval, Canada lynx, and flat-headed cat are specialist predators that prey upon one species more often than others. For instance, the flat-headed cat and the fishing cat prey mostly upon fish and frogs. Indeed, the flat-headed cat does not recognize mice or birds as prey. In a noteworthy experiment, Paul Leyhausen demonstrated that flat-headed cats prey upon aquatic species. Leyhausen introduced a mouse to a captive flat-headed cat. However, the flat-headed cat did not recognize the mouse as a prey item. When the mouse was swimming, however, the flat-headed cat took to the water and pursued its prey. After securing the mouse in its jaws, the flat-headed cat tossed the mouse from the water and, with its paws, rolled the mouse away from water as if the mouse was a frog or fish. The same behavior toward birds was also observed.

Constanza Napolitano, Susan Walker, and their colleagues analyzed scat collected from Andean cat dens in the Andes. They found that Andean cats are specialist predators of mountain viscachas (*Lagidium sp.*), rabbit-like members of the chinchilla family living in large colonies located near water and boulder fields. JGS and his Bolivian colleague, Lilian Villalba, found that in some Bolivian villages in the Andes, humans also eat mountain

A fishing cat attacks and kills a snake. Because this snake is large, the fishing cat will cache the food and return to it for another meal.

Photo © Neville Buck

viscachas. Unlike rabbits, mountain viscachas unfortunately have at most one young per year. Earlier in the twentieth century chinchillas were nearly wiped out by the fur trade and, as a consequence, those Andean cat populations that preyed upon chinchillas disappeared.

Servals are rodent specialists that hunt in the tall grass savannas of Africa. A serval's long ears enable locating rodents by their sounds in the high grass. Once the rodent is located, the serval uses its long legs to leap into the air and land within a swift snatch of its victim. Although this great leaping ability is occasionally used to secure birds, the serval is a rodent specialist.

In Suriname, on the northern coast of South America, Steve Chin A Foeng and JGS found that ocelots were nocturnal predators and, in the same area, jaguarundis were diurnal predators. These small cats occupied the same area but were hunting different prey at different times of the day. It is likely that the ocelot and jaguarundi avoid direct competition simply by avoiding each other.

The leopard cat of Asia is also a nocturnal rodent specialist and is easily observed hunting at night in palm oil plantations, a prime habitat of rats.

The serval uses its large ears rather than its eyesight to locate prey in the tall grass of the African savannas. This mainly rodent hunter employs a number of different methods to flush out a potential meal. Photo © Dale Anderson

The Pallas' cat is a rodent specialist of the Asian steppe. Small but ubiquitous pikas (*Pika sp.*) are a favorite prey.

The Iberian lynx is known to be a European hare specialist. When a contagious virus reduced the hare population of Spain and Portugal by 98%, Iberian lynx populations plummeted. The Iberian lynx became the rarest and most threatened cat in the world. It was the first and only wild cat to be considered Critically Endangered by the International Union for Conservation of Nature. Iberian lynx have not been seen in Portugal since about 1990 and by 2009 the surviving wild population in Spain was estimated at no more than 180 individuals. This stark example illustrates the perils of being a specialized predator.

The Canada lynx is a well-known snowshoe hare specialist. The cat's oversized paws aid its pursuit of its prey through deep snow. Sadly, unlike some big cats, many small cats have never been studied in the wild.

No one is certain what the bay cat preys upon, and we can only guess based on analogies and limited knowledge.

How do small cats hunt?

Small wild cats hunt in ways that are both similar to and differ from their larger relatives. No small cats use pursuit like a cheetah to run down their prey. This is because small cats generally do not prey upon hoofed animals that use their speed to escape. Instead, small cats use stealth to close the distance with their intended victims. The final strike comes swiftly and seemingly out of nowhere, leaving the prey little chance of escape.

Table 7.1. Small cat diet/prey species

Species	Diet/prey species
Asiatic golden cat	Birds, small deer, hares, lizards, and on occasion poultry and livestock
bay cat	Small mammals and birds, small primates, and carrion
marbled cat	Possibly birds. Thought to take squirrels, rats, and frogs
African golden cat	Rodents, hyrax, monkey, birds, and duikers
caracal	Birds, hyrax, rodents, dik-dik, and antelope fawns. Capable of taking prey like impala and reedbuck
serval	Mole rats, small birds, frogs, lizards, and insects
black-footed cat	Gerbils, mice, spiders, and other insects
domestic cat	Birds, mice and rats, lizards, and other small mammals*
jungle cat	Birds, small rodents, lizards, snakes, frogs, and fish. Will tackle larger prey like chital fawns
sand cat	Small mammals, birds, reptiles, and insects like locusts
wildcat	Mainly rodents but also preys on small mammals, small and medium-sized birds, and insects
fishing cat	Fish, snails, snakes, small mammals, and birds
flat-headed cat	Thought to eat frogs, fish, and rodents
leopard cat	Probably hares, rodents, reptiles, birds, and fish
Pallas' cat	Marmots, pikas, small mammals, ground squirrels, hare, and birds
rusty-spotted cat	Possibly eats birds, small mammals, insects, reptiles, and frogs
bobcat	Cottontail rabbits, snowshoe hare, and jackrabbits. Also rodents, opossums, birds, snakes, and deer
Canada lynx	Snowshoe hare, mice, voles, red squirrels, flying squirrels, grouse, and ptarmigan, also caribou fawns
Eurasian lynx	Rabbits and hare. Also mice, birds, and deer. Will take domestic livestock
Iberian lynx	Rabbits, ducks, and fallow deer fawns
Andean cat	Small mammals, birds, and reptiles such as lizards
Geoffroy's cat	Small mammals such as rats, mice, guinea pigs, agouti, and small birds
guigna	Probably feeds on small mammals and birds
margay	Rodents, birds, reptiles, and insects
ocelot	Agouti, birds, fish, snakes, lizards, and land crabs
pampas cat	Ground-dwelling birds, rodents, and wild guinea pigs
tigrina	Possibly eats rodents and small birds
clouded leopard	Small deer, birds, wild pig, monkey, fish, and other small to medium mammals
Sunda clouded leopard	Birds, small mammals, small deer, monkey, young orangutan, and wild pigs
jaguarundi	Small mammals, arthropods, birds, opossums, rabbits, armadillos, monkeys, and also fruit

*Indicates diet/prey species for feral cats and semi-wild domestic cats

Different small cats use different methods. The pampas cat hunting in the Andes walks slowly, stopping frequently to look and listen in different directions before proceeding a few more steps. The flat-headed cat and fishing cat, which prey upon fish and frogs, submerge their heads to find and secure prey. The Andean cat searches for mountain viscachas living in rocky boulder fields. The shrill, high-pitched whistle-like alarm call of

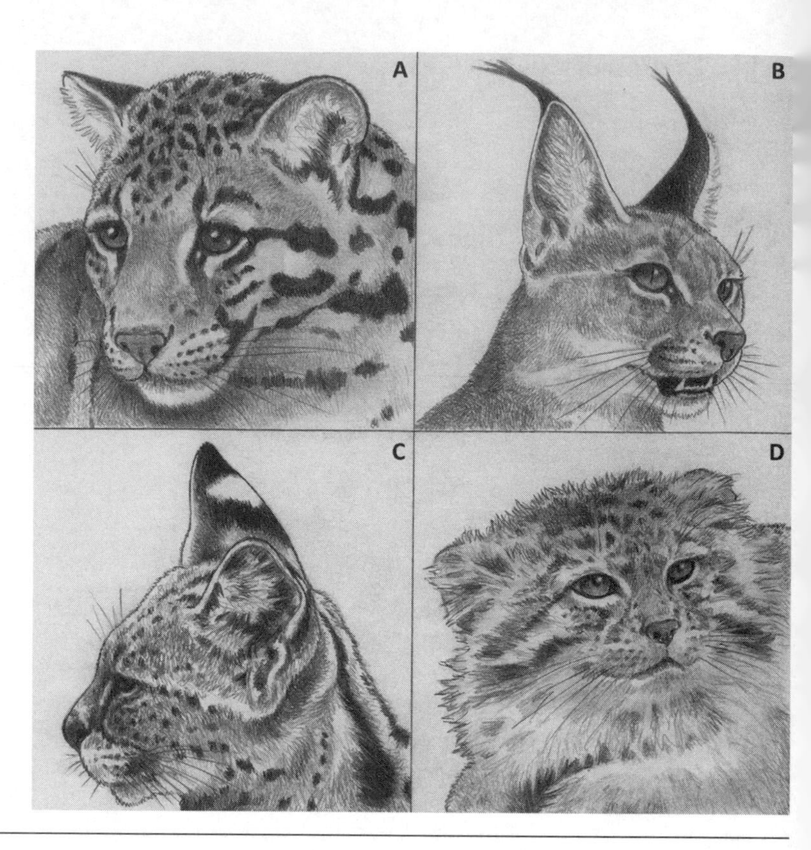

Even among the small cat species, ear size and shape can vary considerably. The four cats illustrated are: *(A)* clouded leopard with its short rounded ears, *(B)* caracal with its tall triangular ears, *(C)* serval with its large oval shaped ears, *(D)* Pallas' cat with its short rounded widespread ears.

the mountain viscacha signals to neighbors that an Andean cat is hunting. Stealthily stalking its prey, using rocks for cover, the Andean cat employs a final lightning-fast attack on its prey; before the mountain viscacha decides which direction to flee, it's too late. The leopard cat of Asia, a nocturnal rodent specialist, slowly walks, stops frequently, and listens for rats scurrying beneath a cover of palm leaves. The serval of the African tall grass savanna is a rodent specialist as well, but uses its large ears and long legs to locate and then leap onto its prey. The caracal and the jungle cat are also leapers and are known to take birds on the wing.

The small rusty-spotted cat of India and Sri Lanka sometimes makes its home very close to people. One of the world's smallest small cats, this beautiful animal is also a nocturnal rodent specialist. Human settlements attract rodents and hence the services provided freely by rusty-spotted cats might be looked upon as beneficial to humans. It's likely that the Pallas' cat of the Asian steppe hunts much as the Andean cat does since both live in similar habitats.

Alex Sliwa has studied one of the smallest small cats in the world, the black-footed cat, in South Africa. The black-footed cat is a nocturnal hunter living in the savanna. Rodents, insects, and birds make up its diet. One night Sliwa observed a black-footed cat attack and kill a bustard, a bird much larger than the cat.

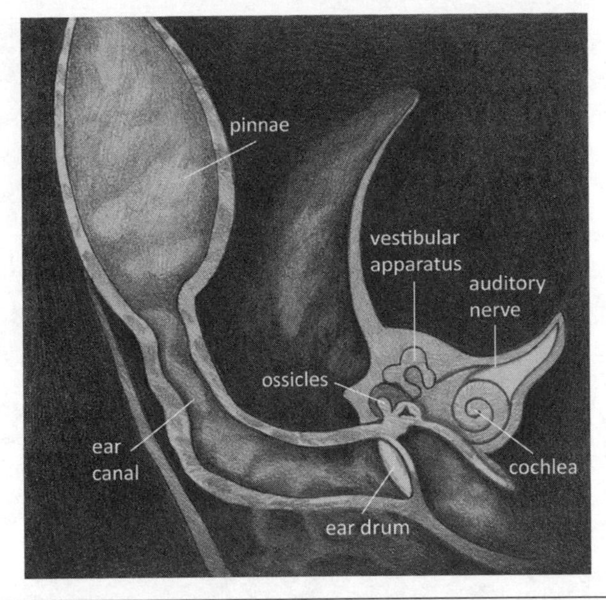

The internal view of the cat's ear is illustrated; not surprisingly it mirrors that of other mammals including humans. Only the cat's ears, or "pinnae," are visible; they are highly mobile and able to accurately pinpoint the direction of sound emitted by prey. Controlled by 30 powerful muscles, they are an invaluable tool and thoroughly exploited by the cat.

The bobcat, Eurasian lynx, and Canada lynx have oversized paws, which more evenly distribute their body weight across the surface of winter snow, and long hair growing between their toes, allowing them to grip the snow while pursuing their prey.

Stealth, complete silence, unlimited patience, and formidable weapons enable small cats to secure their prey. No more formidable predator stalks it prey. The Iberian lynx is a consummate stalker of hares in the brush and rock habitat where it makes a living. When the moment arrives to rush its prey, escape is nearly impossible.

Do small cats hide or bury food?

Small cats generally take smaller prey than themselves, and hence they are able to consume their prey in its entirety. JGS observed an Andean cat weighing 4.5 kg (10 lbs) consume an entire mountain viscacha weighing about 0.8 kg (1.7 lbs); all that remained of the viscacha was a single hind foreleg and paw. The Andean cat consumed the skull, teeth, and entire body including some of the fur. We humans can only imagine what it is like to consume 15% of our body weight at a single meal.

If a large kill is made, a small cat will cache the remains and return to it later to feed. However, as far as is known, no small cats bury food. Generally, animals that dig have exposed claws. Thus, neither big nor small cats are known to dig so they do not bury food. A small cat's claws are designed to grasp prey, not to dig.

A clouded leopard shows its teeth. Clouded leopards are believed to hunt larger prey such as forest deer and primates. Photo © Neville Buck

Rarely are small cats observed when hunting. Here, in the Chilean Andes an Andean cat spots its favorite prey, a mountain viscacha, sitting on a rock. The Andean cat, perfectly camouflaged, is seen crouched on a rock looking at its intended prey that sits on a rock uphill from the cat.

How often do small cats eat and drink?

Small wild cats eat and drink whenever they need to, just like humans. When their appetites are satisfied and their thirst quenched they rest in quiet places, sometimes on the branches of trees, where they are safe from larger predators.

The time between meals is determined by the size of their last meal. Most of what we know about the behavioral aspects of eating is derived from those kept in captivity. We know, for instance, that small cats can go several days without food after a large meal. The sand cat and the black-footed cat can go without water for long periods of time and are presumed to derive moisture from the prey they consume.

However, much of what we believe is true has been extrapolated from a single or a few observations. Scientists believe that smaller small cats consume a higher percentage of their body weight than larger small cats, but this has not been determined quantitatively. Captive guignas have been reported to consume 15–20% of their body weight in meat and bone each day. How much and how often guignas eat in the wild is not known.

Do small cats scavenge?

Probably. Free food is hard to pass up. JGS caught, radio-collared, and released guignas in Chile using packaged chicken parts. To test if guignas preferred live hens to chicken parts, he set two live traps adjacent to each other. The guigna selected the trap baited with chicken parts. Although hens are about the same weight as guignas, the cats are known henhouse raiders. Guignas routinely kill all the hens in a henhouse during a single raid. Nevertheless, given an opportunity to feed on a live hen versus chicken parts, the male guigna chose chicken parts. This is an example of scavenging. The Andean cat that JGS and his colleagues captured, radio-collared, and released in the high Andes of Bolivia was caught using a dead mountain viscacha as bait. This must also be considered scavenging.

Leopard cats have been studied in Borneo where they are well-known to be nocturnal predators of rats. Rats are commonly found in palm oil plantations, and leopard cats prey upon these rats. To capture leopard cats for their study in Borneo, researchers used live rats in small traps. Dead rats failed to attract the attention of the leopard cats, so in this case the leopard cats did not scavenge. Perhaps the leopard cats could not locate their dead prey because it made no noise. It remains to be learned whether leopard cats do in fact scavenge.

Scavenging, the act of feeding on carrion or refuse often carries disgusting connotations. It should not. In fact, humans are scavengers; rarely, if ever, do we eat live food.

Chapter 8

Small Cats and Humans

Do small cats make good pets?

A good pet is one that understands who the boss is, obeys its master, and is not destructive. If you want a pet that can shred a perfectly good sofa to small pieces scattered about the living room, destroy curtains at a single bound, and scratch deep furrows in otherwise lovely table legs and banisters, then a small cat is the perfect pet. Domestic cats make fine house companions but sometimes seem aloof. Small cats will nearly always be aloof. Although no small cat species can be domesticated like a domestic cat, some can be tamed.

The difference in behavior between a tame small cat and a domestic cat is substantial. Even when raised from a kitten, small cats still retain their wildness. A small cat can sometimes be tamed to eat from a bowl and allow itself to be petted, but it will still be more at home in the wild than in its owner's home. Professional animal trainers who work with small cats claim that with several hours or more of *daily* training some small cats can be trained and eventually tamed. Nevertheless, without constant reinforcement a small cat will forget its training and opt for the wild.

Of all wild cats only two have ever been trained to work for humans: the caracal and the cheetah, a big cat. Both were used for sport hunting in India back in the 1800s, when cheetahs still existed in India. The cheetah was used to catch blackbuck, a small Indian antelope, and the caracal's leaping ability was harnessed to hunt birds, sometimes in sport contests.

It was common in the 1800s and early 1900s for natural history writers to comment on whether or not a small cat could make a good house pet. Tamable small cats that were mentioned by early authors include the wildcat, tigrina, guigna, margay, Geoffroy's cat, serval, and clouded leopard. However, tame wild cats should not be confused with a domestic cat. Indeed, even our gentle house cat by day can become an effective nocturnal neighborhood predator. All too many domestic cats have become feral cats, making a living off of wild birds, lizards and frogs, and rodents.

It's not surprising that a rescued small cat kitten seems to make an adorable and interesting house pet, but upon reaching maturity turns downright wild. It would be unusual if this were not the case. We sometimes learn of people who find a small cat kitten that was "abandoned by its mother." Most often the mother is away hunting and is forced to leave the kittens behind. If this happens, by far the best course of action is to resist the temptation to take the kitten home to "save it." Remember, small cats learn from their mother what to eat and how to hunt—skills a human cannot teach. The best course of action is to leave the kittens alone because mom will, in all likelihood, return. If you find a small cat kitten in the wild, do the right thing and leave the kitten alone. To do otherwise is actually a violation of federal wildlife law.

Should people feed small cats?

Although the chance of seeing a small wild cat is exceedingly small, those of us that have been lucky enough never forget the experience. Leaving food out for any wild animal including small cats is not a good idea and should never be done. The food provided by humans is not usually the food needed by a small cat. Moreover, the food might end up being harmful over the long term. The younger the small cat, the more harmful the food will be. We humans who want so much to see a particular small wild cat must resist the temptation of providing food in the hopes of attracting the cat to satisfy our desire to observe them. Small cats do best without human subsidies.

Many small cats benefit for a short time from human subsidics in the form of domestic free-ranging chickens and geese, such as those kept on farms and by people living in rural areas. Eventually, a small cat that becomes a poultry killer is itself the innocent victim of retribution. Many a small cat has lost its life for stealing (in the eyes of a human) chickens.

Do small cats feel pain?

All animals feel pain. The sensation of pain provides important feedback to the brain. The feeling of pain enables the ability to recognize and mitigate or avoid dangerous and potentially fatal external stimulation, whatever its origin. Any organism that fails to feel pain is at great risk and, in all likelihood, will die before it has a chance to reproduce. In this way, the genes that gave rise to the ability to feel no pain will not be passed on to future generations.

Though there is no way to test an animal's degree of pain, it is safe to conclude that small cats and other animals do not feel pain the same way humans do. Certainly, small cats can neither interpret pain the same way humans can, nor anticipate pain before it happens. Certain pain is probably more painful to a small cat than to a human. For instance, excessive stimulation of the vibrissae or high-frequency sounds are likely more painful to small cats than to humans, whose senses are not nearly so well developed. However, most animals, including domestic cats and dogs, endure pain better than humans.

There is no way for us to live in a small cat's skin and experience the world as they do. The difference between small cats and humans is that we can *imagine* what it feels like to be a small cat, but small cats cannot imagine what it feels like to be a human. If, in fact, more of us took the time to imagine what it's like to be a small cat, or any animal for that matter, it's likely that we would not be in the middle of the tragic extinction crisis we are in now.

What do I do if I find an injured or orphaned small cat?

If you are certain, absolutely certain, that a small cat is injured or orphaned, the best course of action is to notify a state wildlife officer. Every state has a game and fish department that is responsible for dealing with wildlife issues. In the United States and in many European countries there are private licensed organizations that also deal with orphaned or injured animals. Although only a small fraction of injured small cats find their way to wildlife rehabilitation facilities, and fewer recover well enough to be released, it's important to make every effort to aid the animals in need of help. However, let professionals do the job because the risk of further injuries during capture is great, and further injury might prove fatal. Remember, an injured small cat or an orphan does not know you are there to help.

Injured small cats have been known to recover from what appear to be near fatal injuries. The old saying "cats have nine lives" has great merit. Domestic cats have been known to fall from windows five or more *stories* high, land on concrete sidewalks, and walk away without injury. There is a good chance that an injured small cat will recover on its own if it can be left in a peaceful, protected place. A wildlife rehabilitation professional can make a determination.

Though not the oldest law, the Endangered Species Act (ESA) is the most widely cited law protecting wildlife in the United States. The ESA became law in 1966, and was amended in 1969, 1973, and 1988. The ESA prohibits importing, exporting, and taking (doing harm to) endangered species. In addition, carrying, delivering, possessing, selling, shipping, and transporting endangered species unlawfully taken in the Unites States is prohibited. The Canada lynx was listed as threatened in the contiguous United States under the ESA on April 21, 2000. Simply transporting an injured or an orphaned Canada lynx can carry a $100,000 fine and a one year jail term in a federal prison. Even mistakenly transporting an injured bobcat that turns out to be a Canada lynx could carry a severe penalty. The best thing to do is notify the state game and fish department at once. The Canada lynx and you will both be better served.

How can I become a better observer of small cats?

Observations of small cats in the wild usually happen randomly and don't last long. That is, most observations are chance occasions that suddenly and unexpected occur, often much to the surprise of the small cat and the observer. However, such observations are never forgotten.

Purposely becoming a better observer of small cats takes more time and patience than most people have to invest. This is true for several reasons. First, because small cats are top predators, they are naturally rare wherever they occur. Second, small cats aren't social. Third, small cats are silent, patient predators. They do not make noise while walking through their habitats, thus you are unlikely to hear them before they hear you. Fourth, small cats are wary of intruders and extraordinary events. Small cats have more acute senses than humans and so will mostly likely sense your approach before you see them. Unlike humans that depend so heavily on their sense of sight, small cats use their eyes, ears, nose, and vibrissae to sense the world. If a small cat senses you first, you will not see it before it makes an almost ghost-like disappearance. Nevertheless, to those with exceptional persistence, it is indeed possible to greatly increase one's chances of observing small cats in the wild. Here is what to do.

Small cats are best observed when they are busy hunting or traveling from one part of their home range to another. Thus, it is essential to be where an individual small cat lives. Learning to recognize favorable habitat is essential. The best chance to see a small cat is to venture into its habitat as far from human areas as you can hike, find a comfortable log to sit on, and wait. If you are moving, the small cat will discover you and take evasive action. If you remain still, this increases greatly your chance of seeing a small cat. There is one last thing you can do to increase your chances of seeing a small cat in the wild: use a lure that the small cat cannot resist. Place a little catnip on a rotten log where you would like to observe the

This Asiatic golden cat kitten appears lost without its mother. However, it is far better off left alone, for there is little doubt its mother will return to it. Adult female small cats must hunt to eat and so are forced to leave the kittens unattended. Should you be lucky enough to find such a kitten, enjoy it from a distance for a short period of time and then leave it alone. The chances are good that its mother will return.

object of your passion. Put some distance between yourself and catnip lure, relax, read a book, and wait. Refresh the lure each day. If in two weeks you are not successful, try a different location.

How do I know whether I have small cats in my backyard?

There are several ways to discover if small cats are in the neighborhood. One way is to look for cat tracks left in soft soil. The problem will be in distinguishing a small wild cat's tracks from a domestic cat's tracks.

A better way to learn if there are small cats in the neighborhood is to use hidden automatic cameras, called trail cameras. The first trail camera was invented by George Shiras III in the late 1890s. Since then, trail cameras have become popular tools for scientists wishing to monitor wildlife such as tigers in India and small cats in Borneo.

Trail cameras typically consist of a digital camera or image recorder and a sensor that detects heat-in-motion. When an animal walks in front of the sensor, heat-in-motion is detected, and the camera is automatically triggered to take a picture. The proper use of a lure on a rotten log helps position the small cat in front of the trail camera. The small cat's picture will be taken automatically when the small cat walks in front of the trail camera.

JGS supplied a trail camera and liquid catnip to his neighbor in rural New Mexico. They suspected a bobcat was patrolling for jackrabbits in the neighborhood. One morning the neighbor heard a commotion outside his bedroom window. Slowly drawing back the window shade, he was stunned and excited to see an adult bobcat hissing and growling at its reflection in the window glass. Repeated pictures from the trail camera provided

These Chileans had several guignas living in their backyard (a very large forest) that were continually preying upon their chickens and geese. JGS was able to capture, radio-collar, and release several of them. These guignas provided the first information on how guignas lived in a human-dominated landscape.

confirmation of the bobcat. Indeed, it's not important what the neighbor *thought* he saw. What is important is the evidence he provided so that JGS could make up his own mind what was observed. The pictures taken by the trail camera provided the proof—hard evidence of the bobcat's presence.

Most sporting good stores carry a wide variety of trail cameras. The old adage "you get what you pay for" applies. A trail camera is well worth the investment, and we have no doubt that you will enjoy the images of wildlife you photograph. Indeed, we doubt you will be able to endure the anticipation and not be able to resist the temptation to *check the set-up* to *make sure it's working*. Who knows: that small cat you had a hunch was stalking about might just show itself in a picture that will provide you with sufficient evidence to convince even your most skeptical friends and colleagues.

Why are small cats important?

Small cats are important to both the ecosystems they live in and the humans that live nearby them. There are no known examples of humans being seriously injured by small cats. Small cats are predators that often take the most common prey species that, without predators, would dominate and perhaps eliminate other species. Thus, small cats maintain biodiversity and the ecological balance in the ecosystems where they live.

Small cats often prey upon rodents that pose direct health threats to humans and damage human's grain crops. Thus, small cats provide an important and often underestimated *ecosystem service* to humans at no cost—they keep rodent populations at bay that would otherwise increase in the absence of predation.

Chapter 9

Small Cat Problems (from a human viewpoint)

Are small cats pests?

They can be. A wild animal is generally considered a pest to humans when it damages human property. Such property is typically food that humans grow to eat or raise for profit. For instance, an elephant is considered a pest when it tramples or consumes a person's garden plants. Small cats are considered pests when they raid a rural henhouse.

Small wild cats are not known to attack humans, even children, so are not considered direct threats. As carnivores, small cats don't damage gardens or crops. However, small cats are considered pests when they steal people's poultry, especially free-ranging chickens, which often prove to be too great a temptation. The larger small cats such as caracal and Eurasian lynx are known to prey upon hoof-stock such as sheep and goats. This makes them hated pests. Retribution is both swift and fatal. Humans are not known for tolerance toward small cats or any other predators doing what predators do naturally and without malice.

Small wild cats are hypercarnivores, obligate meat-eaters. They will take whatever prey they can catch. For small cats, free-ranging chickens are exceptionally easy prey. Small wild cats perform a valuable ecosystem service at no cost to humans—they consume rodents—and so benefit humans and indirectly help protect our grain crops, but a small cat that takes a single chicken is most often considered by humans to be a pest that must be eliminated as soon as possible.

In Chile, guignas are often considered pests, because they prey upon free-ranging chickens and geese. Unfortunately, a guigna that kills chickens becomes a dead guigna. Free-ranging domestic fowl are a widespread serious threat to guignas. Photo © Eduardo and Pilar Rodriquez

How are small cats kept away from people, livestock, and poultry?

Small wild cats generally avoid humans since they do not consider people as prey items, and because dogs, the enemies of small wild cats, are usually associated with people. Thus, small cats generally avoid humans, but sometimes cannot avoid the temptation of easily obtained chickens or geese. Keeping small cats away from chickens is quite simple, but all too often not done.

Small cats are generally good climbers and some like caracals are great leapers. Nevertheless, high chain-link fences with overhangs can be used to keep small cats away from livestock. Guard dogs also can be used to effectively prevent small cats from preying on livestock.

Not surprisingly, one of the biggest threats to small wild cats is the use of free-ranging poultry and the temptation it represents to a hungry cat. The solution to this problem is to restrict the movement of the poultry by using a fish net over the henhouse or pecking ground that prevents small cats from entering while allowing chickens and geese to feed. A fence with an overhang also works.

Small cats are not diggers and will not bite through a net to take chickens. Note that chickens are fed grain that attracts rodents. Rodents experience little difficulty penetrating net defenses. Thus, the protected rodent population might increase and become a serious threat, not just to the chickens, but to humans as well. JGS recalls a henhouse in Chile being so rodent infested that even the chickens declined to enter the shelter. The chicken feed, grain, attracted rodents. The chickens and rodents attracted

A high fence acts as a deterrent to small cats preying on domestic fowl. However, even the diminutive guigna can climb a fence. Domestic dogs are enough to frighten some small cats.

guignas. Guignas will take whichever is easiest to catch—chickens. If the guignas are kept at bay, then a more serious problem, a rodent infestation, can result. Rodents carry diseases that can be transmitted to humans.

Are small cats vectors of human disease?

Small cats transmit diseases between themselves, but disease transmission to humans is rare. An infectious disease that is transmitted from a small cat or domestic cat (the vector) directly to a human is called a *zoonosis*. Zoonoses are caused by bacterias, fungi, parasites, and viruses.

Transmission of an infectious disease from a small cat to a human is exceedingly rare simply because small wild cats avoid humans. For instance, rabies can be transmitted by a domestic cat to its human owner, or to a human by a small cat with rabies that bites a human. Rabies transmission from domestic cats to humans is rare, and from small wild cats to humans, though theoretically possible, is exceedinlgy rare.

Small cats and domestic cats are more likely to be indirect delievers of diseases to humans. For instance, in November 2007, a biologist studying pumas (a big cat) in Grand Canyon National Park, Arizona, died several days after contracting plague from a flea that was found on a dead puma he was examining. Hanta virus can be fatal to humans and is transmitted by rodents that might be delivered to a doorstep by a domestic cat. Thus, while not directly transmitting diseases, small cats can act indirecly to transmit diseases. However, the chance of a small cat, not a domestic cat, leaving an edible prey item such as a disease-ladened rodent at a human's doorstep must be considered exceedingly rare.

Though humans and domestic cats share some diseases such as skin problems, it is rare that a disease passes directly from a domestic cat to a human. Because small wild cats avoid interations with humans, disease transmission is exceedling rare. The benefit of having small cats in the area far exceeds the potential risks.

Chapter 10

Human Problems (from a small cat's viewpoint)

Are small cats endangered?

Some small cats are and some are not endangered. In its common use the word "endangered" is a catchall term that refers to the conservation status of a species. Some take the term to mean that a species is "threatened with extinction" while others mean that a species' population is declining. To those of us charged with assigning the conservation status of each species of small cat, the word "endangered" has a very specific meaning. Understanding and using words like endangered, threatened, and least concern properly is the first step in establishing a relative ranking of conservation priorities.

The world's foremost organization dedicated to the conservation of nature is the International Union for Conservation of Nature, or IUCN. Founded in 1948, the IUCN's headquarters is located in Gland, Switzerland. More than 80 countries, about 800 non-government organizations, and thousands of volunteer scientists and experts comprise the IUCN. The IUCN's mission is to

> influence, encourage and assist societies throughout the world to conserve the integrity and diversity of nature and to ensure that any use of natural resources is equitable and ecologically sustainable.

The IUCN has three components: member organizations, six scientific commissions, and a professional secretariat.

One of the six scientific commissions and by far the largest is the Species Survival Commission (SSC). The SSC advises the IUCN on the technical aspects of species conservation and mobilizes action for those species that are threatened with extinction. The SSC charges each of its specialist groups with contributing to the IUCN Red List of Threatened Species.

The IUCN Red List is the world's most comprehensive inventory of the global conservation status of plant and animal species. The Red List uses a set of criteria to evaluate the extinction risk of thousands of species and subspecies, including small cats. These criteria are relevant to all species and all countries. The IUCN Red List is recognized as the most authoritative guide to the status of biological diversity. The Cat Specialist Group of the SSC is responsible for assigning all 36 species of wild cats, big and small, a ranking representing its conservation status.

The goal of the Red List is to call attention to the scale of conservation problems to the public and to policymakers and to motivate the global community to reduce the risk of species extinctions. In principle, the Red List should be used by government agencies, wildlife departments, conservation-related non-governmental organizations (NGOs), natural resource planners, educational organizations, and many others interested in halting what is now commonly referred to as the *Sixth Mass Extinction*, the greatest loss of biodiversity the world has experienced in the last 65 million years.

There are nine categories in the IUCN Red List system: Extinct, Extinct in the Wild, Critically Endangered, Endangered, Vulnerable, Near Threatened, Least Concern, Data Deficient, and Not Evaluated. The words *threatened species* specifically refers to those species listed as Critically Endangered, Endangered, and Vulnerable.

Classification into the categories for threatened species is through a set of five quantitative criteria. These criteria are based on biological factors related to extinction risk and include the estimated rate of population decline, estimated population size, area or geographic distribution where the species occurs, and degree of habitat fragmentation within the occupied area. This information is supplied by IUCN's SSC, a network of thousands of experts on plants, animals, and conservation issues, as well as by partner organizations.

Revisions of the Red List are produced periodically. The 1996 Red List revealed that one in four mammal species and one in eight bird species faced extinction; the 2002 Red List confirmed that the global extinction crisis was as bad as or worse than believed. Dramatic declines in populations of many species, including reptiles and primates, were reported. A new assessment for wild cats was completed in 2007, and in late 2008 a new wild cat Red List was made public.

As of mid-2009, the Iberian lynx, a small cat, was the only wild cat listed as Critically Endangered. Six wild cats, including four small cats, were listed as Endangered, nine wild cats including seven small cats were listed as Vulnerable, nine wild cats including seven small cats were listed as Near

Table 10.1. Conservation status of small cats globally

Critically endangered
Iberian lynx *(Lynx pardinus)*, Portugal, Spain

Endangered
Andean cat *(Leopardus jacobita)*, Argentina, Bolivia, Chile, Peru
bay cat *(Catopuma badia)*, Borneo
fishing cat *(Prionailurus viverrinus)*, Southeast Asia
flat-headed cat *(Prionailurus planiceps)*, Malaysia, Sumatra, Borneo, Thailand

Vulnerable
black-footed cat *(Felis nigripes)*, southern Africa
clouded leopard *(Neofelis nebulosa)*, Southeast Asia, southern China
guigna *(Leopardus guigna)*, Argentina, Chile
marbled cat *(Pardofelis marmorata)*, Southeast Asia
rusty-spotted cat *(Prionailurus rubiginosus)*, India, Sri Lanka
Sunda clouded leopard *(Neofelis diardi)*, Sumatra, Borneo
tigrina *(Leopardus tigrinus)*, Central and South America

Near threatened
African golden cat *(Caracal aurata)*, Central Africa
Asiatic golden cat *(Catopuma temminckii)*, Southeast Asia
Geoffroy's cat *(Leopardus geoffroyi)*, South America
margay *(Leopardus wiedii)*, South America
Pallas' cat *(Otocolobus manul)*, Asia
pampas cat *(Leopardus colocolo)*, South America
sand cat *(Felis margarita)*, North Africa, Middle East

Least concern
bobcat *(Lynx rufus)*, North America
Canada lynx *(Lynx canadensis)*, North America
caracal *(Caracal caracal)*, Africa, Asia
Eurasian lynx *(Lynx lynx)*, Europe, Asia
jaguarundi *(Puma yagouaroundi)*
jungle cat *(Felis chaus)*, Egypt, Middle East, Southeast Asia
leopard cat *(Prionailurus bengalensis)*, Southeast Asia, China, Taiwan, Korea
ocelot *(Leopardus pardalis)*, Central and South America
serval *(Leptailurus serval)*, Africa
wildcat *(Felis silvestris)*, Africa, Asia, Europe

Source: Red List – The International Union for Conservation of Nature (IUCN)

The Andean cat is the only IUCN Red List Endangered cat in the Americas. Authorities believe less than 2,500 individuals survive.

Threatened, and eleven wild cats including ten small cats were listed as Least Concern (see table 10.1). In summary, 12 of the 16 most threatened wild cat species are small cats.

Will small cats be affected by global warming?

Yes. Global warming is a significant threat to most small cats. As always, global warming, like many other causes of species' loss, will affect some species more seriously than others. Nevertheless, the threat of extinction has significantly increased because of global warming. For instance, the Andean cat is particularly vulnerable to global warming.

The Andean cat lives high above the tree line in the highest and driest areas of the Andes of South America. The Andean cat's prey is small mammals such as the mountain viscacha, which look very much like a common North American cottontail rabbit. Mountain viscachas eat coarse spikegrass that is found near small mountain streams. These streams are glacially fed. Global warming is causing the glaciers to shrink. Glaciers are melting more rapidly than they can accumulate ice in winter. In South America, glaciers are shrinking faster than in other parts of the world. With the loss of glaciers, mountain streams will dry. As a consequence of the loss of mountain streams, grass that the mountain viscachas feed upon will disappear, and so will the mountain viscachas. Without prey, the Andean cat cannot survive. To make matters worse, there are no Andean cats in any zoos in the world.

The glacier on Vulcan Chungara in Chile provides water for the Andean cat's prey, mountain viscachas. As glaciers recede and disappear, freshwater streams also disappear. The loss of streams will lead to the loss of grass upon which mountain viscachas feed. Mountain viscachas and Andean cats will also then disappear. Experts agree and evidence suggests that some glaciers will in fact disappear.

Global warming will have far-reaching impacts on human food production. New areas not formerly under cultivation will need to be brought into production. This implies an increase in habitat fragmentation and loss in our natural ecosystems such as the Amazon forest.

Coastal lowland areas are also expected to be adversely impacted by rising seas levels. Endangered small cats like the flat-headed cat and fishing cat are known to hunt for food along coastal estuaries that might be altered or changed entirely.

Global warming is expected to have greater impacts in high latitudes and high elevations than elsewhere. The impact on less threatened species occupying high latitudes such as the Canada lynx, Eurasian lynx, and Pallas' cat has not yet received the attention it deserves.

Are small cats ever invasive species?

The domestic species is considered an invasive species in places where small cats did not naturally occur. Invasive species are those species that thrive in a human-dominated landscape and are often transported by humans from one place where the species naturally occurred to another where they do not naturally occur. Some species in new lands find themselves without competitors, and so their populations expand rapidly, most often at the expense of one or more native species. When this happens, these introduced, exotic, or invasive species become pests that can threaten other species. Some island bird species have been driven to extinction by human introduced species. Thanks to the help of humans, one small cat has in fact driven other species to extinction.

Mountain viscachas depend on glacial fresh water that enables the grass on which they feed to grow.

In the Felidae, the family of cats, only one species has successfully expanded its geographic range, which now includes continents that throughout the entire history of life on earth have never felt the gentle touch of a cat's paw. Dominant big predators such as the lion and tiger, whose ranges contracted dramatically during historic times, have been unable to do what descendants of a small wildcat originally from the Middle East of Asia has achieved. By forming a mutualistic relationship with humans in the Middle East, the so-called cradle of civilization, future generations of the wildcat became what we now call the domestic cat, which has been spread across the earth, reaching places no wild cat could.

Now, some domestic cats have returned to a more or less wild state, living on their own even in urban landscapes. These domestic cats gone wild are referred to as *feral* cats. In Australia, feral cats are successfully out-competing native species and in some places threatening native species. Spread to remote islands by humans, domestic cats have caused the extinction of some bird species. In parts of the United States feral cats have become a nuisance, a highly successful invasive species.

The domestic cat, however, is the single exception within the Felidae. All other species have seen their geographic ranges reduced, sometimes significantly, by humans even though most species benefit humans by preying upon rodents and other crop-destroying species.

Do people hunt and eat small cats?

Yes, in some places. Aside from retaliation killing of small cats for taking poultry or hoof stock, small cats are hunted for a variety of reasons.

Native Americans living in the Andes attribute supernatural powers to cats. Cats, when killed, dried, and decorated are called "*titi*." When displayed titi are believed to bring good luck. In this tiny store in the Bolivian Andes, JGS stands below a titi that is openly displayed.

Amerindians living in the high Andes kill Andean cats and pampas cats for the supernatural powers the people believe the cats possess. To harness these powers the small cats must be killed, dried, and decorated with dyed wool and silver coins. The decorated dead cat, called a "titi," is then openly displayed as a symbol of good luck.

In some countries, ocelots are hunted to make household accessories such as seat coves. For instance, in shops in Bolivia, Peru, and Ecuador skins of ocelots are often sold in storefront windows. Cat skins from Bolivia were advertised with other products in the centerfold of an Air France Magazine found in the seat pocket on routine flights between Paris and La Paz, the capital city of Bolivia.

Amerindians living in the forests of Central and South America hunt animals, including small cats, for food. Because the skins of several species of small cats are considered to be especially beautiful, these might be made into clothes.

Gun ownership is not prohibited in most Central and South American countries. Laws regarding hunting are often overlooked or ignored in many countries. Hunters are not selective, and anything that moves is considered fair game. Such hunters show no regard for any wildlife and will kill an adult or young of any species, including a small cat. Fortunately, small cats are extremely shy and are difficult to find.

In many parts of Africa, all animals are hunted for food by wide variety of methods. Traps are particularly dangerous because they are not selective and catch whatever steps in them. Whatever is caught ends up being eaten. In Southeast Asia traps are commonly used in great numbers to secure food.

Small cat skins are indeed beautiful. This ocelot skin is for sale in Ecuador. Small cat skins are offered openly for sale, even in countries that are signatories to the Convention on International Trade in Endangered Species (CITES). CITES is an agreement, not a law, and cannot be enforced.

The natives of Borneo such as the Iban and Dayak are skilled hunters that use blowpipes to kill whatever they find in the forest. Nothing escapes these exceptionally talented hunters. Even the rarest and most threatened small cats such as the bay cat are not spared and end up in the cooking pot. The people simply do not realize they are killing and eating an IUCN Red List Endangered cat and perhaps the rarest cat on earth.

In Mongolia, the local pastoralists hunt and kill Pallas' cats for their skins, which are made into accessories like gloves and coat lining. More than 30 Pallas' cat skins are required to make one winter coat. This is also true in China where Tibetan pastoralists kill small cats such as the Chinese mountain cat to make clothing accessories and for the fur trade. On the streets of several large cities in China, vests, scarves, gloves, and coats made of small cat skins can be openly purchased.

Globally, illegal killing of small cats and all wildlife continues to take an unacceptably heavy toll.

Why are some small cat skins so valuable?

The fur of several species of small cats is considered to be exceptionally attractive to humans, is therefore sought after, and consequently commands a high price compared to other alternatives. This explains why fur coats remain both popular and expensive. Though international trade in threatened species is prohibited by the Convention on International Trade in Endangered Species (CITES), local markets continue a brisk trade. Indeed, so-called *skin shops* in many countries continue to offer small cat skins for sale, especially to tourists, who then assume the risk of transporting them internationally. Such risk carries with it the possibility of fines and even imprisonment.

Why do small cats get hit by cars?

The demands of small cats are relatively simple to appreciate. All small cats require prey to sustain themselves and adequate space to raise their young. Small cats seek nothing from humans except to be left alone to live out their lives. But in today's world space and food are increasingly difficult to find.

The human population on earth is approaching 7,000,000,000,000 (7 billion). Humans are the most successful mammalian species in the history of life on earth. In addition, mammalian species such as cattle, sheep, goats, pigs, horses, domestic dogs and cats, which enjoy the full protection and care offered by humans, have greatly increased in numbers as well. Other species that humans make use of, such as chickens and geese, have also multiplied. All these species require food and space.

Natural landscapes have been and continue to become increasingly more fragmented. With habitat fragmentation have come more roads. With roads comes increased traffic into more and more remote areas. As the number of humans increases, more natural landscapes must be converted to food production. More roads are built to accommodate increased traffic. It's no surprise that small cats die as a result of being hit by cars and trucks.

Even though small cats do not harm people directly, their numbers continue to decline. Unless we humans make room for small cats, they too will pass into extinction before they are given a chance to evolve into new species.

Jungle cat *(Felis chaus)*, Israel.

Photo © Alex Sliwa

Leopard cat *(Prionailurus bengalensis)*, Sabah, Malaysian Borneo.

Photo © Jim Sanderson

Marbled cat *(Pardofelis marmorata)*, Sumatra. Photo © Sugesti Mhd. Arif, Abu Lubis, and Jim Sanderson

Margay *(Leopardus wiedii)*, captive, Brazil. Photo © Jim Sanderson

Ocelot *(Leopardus pardalis)*, Brazil.
Photo © Alex Sliwa

Pallas' cat *(Otocolobus manul)*.
Photo © Neville Buck

Rusty-spotted cat *(Prionailurus rubiginosus)*, Sri Lanka. Photo © Jim Sanderson

Sand cat *(Felis margarita)*, captive, USA. Photo © Jim Sanderson

Serval *(Leptailurus serval)*, captive, Australia. Photo © Patrick Watson

Sunda clouded leopard *(Neofelis diardi)*, Sabah, Malaysian Borneo. Photo © J. Ross and A. J. Hearn

Tigrina *(Leopardus tigrinus)*, captive, Brazil. Photo © Jim Sanderson

Wildcat / Chinese mountain cat *(Felis bieti)*, China. Photo © Jim Sanderson, Yin Yufeng, Drubyal Tahksang

Pampas cat *(Leopardus colocolo)*, captive, Chile. Photo © Roberto Cruz

Domestic cat / Maine coon *(Felis catus)*, Australia. Photo © Patrick Watson

Domestic cat / Burmese *(Felis catus)*, Australia. Photo © Patrick Watson

Domestic cat / Short hair *(Felis catus)*, England. Photo © Patrick Watson

Chapter 11

Small Cats in Stories and Literature

What roles do small cats play in religion and mythology?

Small cats have played a big role in religion and mythology. Throughout most of human history, humans were hunter-gatherers. Around 15,000 years ago, at a place now referred to as the Fertile Crescent in Middle East, humans slowly began changing their basic way of securing food. Traditional hunter-gatherers hunted animals and harvested wild vegetables, fruits and nuts, and grains. Probably because of the uncertain and highly variable nature of their food supply, or perhaps because of increased competition from other groups of hunter-gatherers, small groups of people began planting crops, cultivating gardens, and domesticating animals. About this time, wild cats started preying upon rodents attracted to human's stored food supplies. The Middle Eastern wildcat earned its living eating human pests like rodents and killing venomous snakes, probably at night and out of sight of people.

Most likely, humans ignored the small cats, or perhaps even welcomed them with something to drink or eat. Big cats such as the leopard, lion, and cheetah, which also inhabited the region long ago, were exterminated because of their obvious threat to humans and their livestock. But the much smaller wildcat was different. The wildcat was not threatening to humans or their livestock. The wildcat was allowed to come and go as it pleased because it performed a valuable free service that benefitted humans. From this early beginning, a complex relationship between humans and small cats developed. This complex relationship continues today.

Small cats have played a role in religion and mythology from the dawn of recorded history to the present day. Arguably, no other wild animal has played so important a role in human affairs as the small cat. The relationship between small cats and the citizens of ancient Egypt has fascinated both scientists and cat aficionados alike. Small cats have been held in high esteem by some religions and with great contempt by others.

Written stories of small cats and humans date back many thousands of years. The Greek historian Herodotus, who lived in the fifth century BC, is considered the father of written history. Herodotus authored a book known today as *The Histories*, in which he recorded many of his personal experiences, gleaned from his travels throughout the region, and stories that were told to him that he believed to be factual. In one of his books, Herodotus wrote that nothing was more remarkable than the respect paid by the Egyptians to their sacred animals. However, with a single exception, not all individuals of a sacred species were held in high esteem. Only those individuals of the sacred species that lived at temples and shrines were considered special. But one species was different from all the others. Without exception, every single small cat throughout Egypt was considered sacred. Neither wild nor tame dogs ever enjoyed esteem even close to that afforded to small cats. When a cat died, every member of the household mourned and some shaved their eyebrows out of respect. If a house was burning, Herodotus wrote, cats, like other members of the household had to be saved.

Later, the first century BC Greek historian, Diodorus Siculus, wrote his *Historical Library*, a collection of forty books. Book I describes the history and culture of Egypt during this period. He wrote that around 60 BC a Roman soldier in Egypt killed a cat quite accidentally when the wheel of his chariot rolled over it. Witnesses were so infuriated by the incident that, although at the time Romans ruled Egypt, nothing would placate the crowd except the soldier's death. The pharaoh Ptolemy XII tried to explain that the cat was killed accidentally, but nothing would quiet the crowd. Diodorus Siculus claims to have witnessed the offending soldier's execution.

Reverence for small cats had a long history in Egypt. Egyptians held that their gods assumed bodies of animals living around them. Like sports teams and fans today, cults were formed around the sacred animals. For thousands of years before Greek historians recorded the respect held by Egyptians for cats, Egyptians believed cats were deities—embodiments of one of their gods. In early Egyptian mythology, Mafdet was depicted as a woman with the head of a lion—a lion-headed goddess.

Later, another feline-headed goddess, Bastet, rose to prominence. Perhaps as a result of the extinction of lions in the region, Bastet is depicted with a wildcat or domestic cat head. No other small cat has been more closely associated with any other culture at any time in history. Bastet was the deity representing protection, motherhood, and fertility. Not surprisingly, the animal that exhibits these qualities without inhibition is the domestic cat. All small cats were believed to be the flesh and bone embodiment of the deity Bastet, and were thus afforded all rights given to humans.

As a revered animal and one very important to Egyptian society and religion, the small cat was given the same mummification after death as humans. Mummified cats were given in offering to Bastet. In 1888, an Egyptian farmer accidentally uncovered a large tomb containing tens of thousands of mummified cats and kittens. Just how sacred were small cats? This discovery outside the present day town of Beni Hasan contained around eighty thousand cat mummies dating back to 1000–2000 BC. That's how sacred small cats were!

Herodotus wrote that the annual festival of Bastet, held in Bubastis, attracted more than 700,000 people from all over Egypt and the Mediterranean. Eventually the rulers of Bubastis became the rulers of Egypt, and the importance of Bastet spread throughout the empire. Herodotus visited and described the shrine, or temple, in honor of Bastet that was located near the center of the city of Bubastis. He wrote that although the shrine of Bastet was not as large as shrines in other cities, none was more attractive. Cats that died anywhere in Egypt, Herodotus wrote, were often taken to Bubastis to be mummified and buried. At the shrine of Bastet in Bubastis many thousands of mummified cats were later excavated.

Herodotus described the temple and its setting in great detail. The shrine was surrounded by a moat and so appeared to be an island. The shrine of Bastet occupied the center of the island. The island was populated by a vast number of small cats, all considered sacred. Surely this explains, at least in part, why so many cat mummies were found at the site. Without question there must have been some sort of local population control, or perhaps disease and starvation regularly reduced the population.

As with any such event even today, Bubastis also attracted peddlers and artisans, who sold their wears to travelers. One can only imagine travelers from across the region arriving with the remains of their house cats, paying to have them mummified, leaving them at the shrine, and shopping for bronze sculptures, amulets, talismans, artwork, pottery, and carvings—all one way or another depicting small cats and Bastet. It must have been a truly spectacular gathering!

One of the most fascinating artifacts from this period is a bronze statue of a seated cat that represents the goddess Bastet. Known for its collector, Major Robert Grenville Gayer-Anderson, the statue is housed at the British Museum. A copy is also displayed in the Gayer-Anderson Museum in Cairo. The statue, about 42 cm (16.5 in) high and 13 cm (5.2 in) wide, has been thoroughly studied, and scholarly papers have been written about it.

The Gayer-Anderson cat at the British Museum was from Egypt and later donated to the museum. It has been dated at about 600 BC. Photo © Google images

The cat is adorned with jewelry and a protective Wedjat amulet. A winged scarab appears on the chest and also on the head of the cat. Both eye sockets clearly held some sort of inlaid precious or semiprecious stone that long ago fell out and were lost. The statue is now dark green bronze color that is the result of polishing. The Grayer-Anderson cat dates from 660–330 BC and was acquired in the late 1800s.

In Egypt around 400 AD, the cult of Bastet was officially banned by the ruling party. Although small cats lost their sacred status, they never lost their usefulness to humans.

Small cats and religions. Reverence for small cats was not unique to Egypt. The Prophet Muhammad was said to have loved his pet small cat Muezza, which he thoughtfully cared for. Muhammad was likely kind to all animals, and Muslims likely treated small cats kindly. The same cannot be said for dogs, which were and still are considered ritually impure, so dirty that an angel will not enter a house with a dog, or even a picture of a dog. Hindus are known to respect all life, including small cats. This is in sharp contrast to early Christianity, whose followers believed that small cats ranked somewhere close to the devil. However, over time, small cats

Cats were mummified and placed at the shrine of Bastet in Egypt during the annual festival. Many tens of thousands of cat mummies have been unearthed, providing strong evidence that all cats were considered sacred in ancient Egypt. Photo © Google images

were tolerated wherever people lived, including Christian churches, because cats kept rodent populations at bay.

In 1892, St. George Mivart wrote:

> Pope Gregory the Great, who lived towards the end of the sixth century, is said to have had a pet cat, and cats were often occupants of nunneries in the Middle Ages. The great value set upon the cat at this period is shown by the laws which in Wales, Switzerland, and Saxony, and other European countries, imposed a heavy fine on cat-killers. As compensation, a payment was required of as much wheat as was needed to form a pile sufficient to cover the cat to the tip of its tail when held vertically with its nose resting on the ground.

Nine lives of cats. Who among us doubts that cats have nine lives? No one knows where this myth originated. Perhaps the origin is ancient Egypt, where most people kept these sacred animals and saw first hand their unique abilities. Had the expression not been known, it surely would have been coined with the arrival of multi-story buildings with windows and patios that open to the outdoors. No other similar or larger-sized animal, especially humans, has the ability to survive a fall from such high places. Indeed, it seems that the higher the fall, the more likely it is that a cat will survive. This is in stark contrast to what happens to other animals, including humans.

In the video *A Cat's Nine Lives*, presented by National Geographic Video, a small cat's ability to survive falls that fatally injure other animals is scientifically investigated. Cats land on their feet because they have an

A cat's righting reflex or ability to position its body to land on all four limbs is clearly illustrated opposite. Employing its sense of balance and flexibility, an adult cat will twist its body through a series of movements until its feet are facing downward. Once righted it slows its descent by spreading its legs and at the point of contact flex its legs to reduce the force of impact.

innate ability to execute a complex twisting motion that, seen in slow motion, elicits envy from the most talented gold-medalist Olympic diver.

The life of a predator is not an easy one. Prey animals do not give in easily. Often, small cats sustain injuries, often terrible injuries, during the capture of prey. However, small cats are able to go for many days, even a few weeks, without food and water. If a secure and sheltered place can be found to rest, a small cat has an increased chance of surviving an injury that would result in a fatality to almost any other creature. Scientists discovered that the frequency of a cat's purr can heal broken bones faster. Indeed, small cats are masters at resting and purring—two abilities essential to recover from injuries.

Are some small cats considered bad luck?

In the Middle Ages in Europe, Christianity and the church had a powerful grip on the people. Perhaps because of their nocturnal habits, cats and many other animals were associated with the devil, the enemy of all that was Christian. Witches were believed to keep cats, especially black cats. Unlike in ancient Egypt, a popular pastime among Christian followers was torturing and killing cats. The cat's death scream was believed to be the screams of Lucifer.

Domestic cats were also used in animal sacrifices during this time. Perhaps this explains why the Black Plague was so virulent. Without cats to keep rodent populations in check, the plague spread far and wide. Fortunately, those awful days are behind us. However, even in our modern world some people believe that black cats are bad luck.

While working on Chiloé Island, Chile, JGS discovered that local residents believed black cats were especially bad luck. Black kittens were killed at birth. Guignas have two color morphs: a black-spotted form and a melanistic (all black) form. Local Chilótes (the people of Chiloé) associated guignas with very bad luck and referred to them as vampires that hunted at night, sucked the blood of their victims, and dwelled in dark caves by the sea. One Chilóte told JGS he could study guignas on her property only if he agreed to exterminate all the guignas when his study ended. This is a 2.2 kg (5.5 lb) small cat!

In contrast, Native Americans living in the high Andes of Argentina, Bolivia, Chile, and Peru consider the Andean cat extremely powerful good luck. Unfortunately, the Andean cat must be killed to harness this good luck.

In Japan, Maneki Neko translates as "lucky cat." Maneki Neko is the cat that is often seen at the entrance to shops with its warm smile and raised paw inviting passersby to enter.

Working in Thailand, Lon Grasman discovered that Asiatic golden cat hair is considered good luck and to protect the lives of those who have it.

What roles do small cats play in popular culture?

Every cat aficionado probably has a favorite cat character. Many are unforgettable, even in countries far from where the characters were created. Who can forget Dr. Seuss' *The Cat in the Hat*? Small cats have been and remain figures in popular culture transcending international boundaries. There is even a museum known as the Cat Museum in Kuching, Sarawak, on the island of Borneo, which is dedicated to everything cats. Kuching, the capital city of Sarawak, means cat in the Malay language. Los Gatos, California, whose name translates to "the cats" has nothing on Kuching. Kuching features statues of cats at major road intersections.

In the 1920s, long before movies included sound, Felix the Cat was a well-known cartoon character. He had a black coat, white eyes, and a giant

At the Cat Museum in Kuching, Sarawak, Maneki Neko raises its right paw in welcome. Maneki Neko is the lucky cat of Japan.

An Andean cat *titi* in Bolivia is considered good luck.

tooth-filled uncat-like grin. Felix the Cat was the first cartoon character in the world to gain worldwide recognition. He even had his own newspaper comic strip. In the television show, Felix the Cat could assume all sorts of shapes, and his body knew no bounds on flexibility. However, no matter what shape Felix the Cat was stretched into, he remained sneaky and mischievous and was always instigating trouble.

Garfield, the unforgettable and insatiable orange tabby created by Jim Davis, has enjoyed immense success since he appeared in 1978. Garfield is featured in a number of venues including a comic strip, animated televi-

Intersection in Kuching, the capital city of Sarawak, features cats. Two domestic cat adults guard their kittens that are playing in a fountain. In the Malay language "kuching" means cat.

The Cat Museum in Kuching, Sarawak, is the only museum on earth dedicated to cats. If you love cats, plan on spending half a day at the museum, reachable by taxi on a hill outside the capital city of Kuching.

sion specials and series, and big screen movies. Garfield is also featured on popular merchandise. Able to transcend a generation, Garfield retains his place in newspaper comic strips across the United States.

Tom and Jerry, the domestic cat and mouse cartoon duo, was created by William Hanna and Joseph Barbera for Metro-Goldwyn-Mayer. Tom, perhaps short for tomcat, and Jerry, a house mouse, fought never-ending battles. Today these battles would probably be considered far too violent for children's cartoons. Moreover, the cartoon lacked political correctness that was later edited out on reruns. But in the 1950s Tom and Jerry

A Japanese artist has dressed kittens and placed them in a pool hall.

entertained children across the United States who would later be known as the baby boomers.

Sylvester the Cat was a black and white cat with a very bad lisp. Just as Tom's nemesis was Jerry, Sylvester's nemesis was Tweety Bird, a yellow canary with bright orange feet, who also had a speech impediment. Sylvester the Cat and Tweety Bird appeared in Warner Brothers *Looney Tunes* and *Merrie Melodies*. Some cartoons opening with Tweety Bird singing in his cage. Sylvester quietly tip-toed through the room and peered into the cage. Tweety turned and faced the audience, and said "I tot I taw a puddy tat." Tweety looked at Sylvester, then the audience, and said "I did, I did, I did, tee a puddy tat."

Japanese artists have photographed small cats dressed as various characters in interesting scenes. The cats are typically dressed in outfits to make them look like people. They are placed in scenes, often with other dressed cats, that are typical in our everyday experiences.

Are there popular sayings about small cats?

As might be expected, most popular sayings about small cats involve domestic cats. None would make sense if "Andean cat" or "Pallas' cat" was substituted for domestic cat. To understand why, try substituting flat-headed cat in place of the word cat in some of my favorite stories and sayings below.

The domestic cat—the lion of mice.

Carl von Linné (Linneaus) in the first scientific description of the domestic cat, 1758

Lettin' the cat outta the bag is a whole lot easier 'n puttin' it back in.

Will Rogers

Many superstitious people believe that if a black cat crosses their path, bad luck will soon befall them. Groucho Marx dispelled this belief.

A black cat crossing your path signifies that the animal is going somewhere.

Groucho Marx

If a dog jumps in your lap, it is because he is fond of you; but if a cat does the same thing, it is because your lap is warmer.

Alfred North Whitehead

Cats regard people as warm-blooded furniture.

Jacquelyn Mitchard, *The Deep End of the Ocean*

Cats are smarter than dogs. You can't get eight cats to pull a sled through snow.

Sir Winston Churchill

Operationally, God is beginning to resemble not a ruler but the last fading smile of a cosmic Cheshire cat.

Sir Julian Huxley

If you are worthy of its affection, a cat will be your friend but never your slave.

It is difficult to obtain the friendship of a cat. It is a philosophical animal . . . one that does not place its affections thoughtlessly.

Théophile Gautier

As an inspiration to the author, I do not think the cat can be over-estimated. He suggests so much grace, power, beauty, motion, mysticism. I do not wonder that many writers love cats; I am only surprised that all do not.

Carl Van Vechten

As anyone who has ever been around a cat for any length of time well knows, cats have enormous patience with the limitations of the human kind.

Cleveland Amory

> "But I don't want to go among mad people," said Alice. "Oh, you can't help that," said the [Cheshire] cat. "We're all mad here."
>
> LEWIS CARROLL

> I believe cats to be spirits come to earth. A cat, I am sure, could walk on a cloud without coming through.
>
> JULES VERNE

> I care not much for a man's religion whose dog and cat are not the better for it.
>
> ABRAHAM LINCOLN

> I have no desire to be a cat, which walks so lightly that it never creates a disturbance.
>
> ANDREW TAYLOR STILL

> In nine lifetimes, you'll never know as much about your cat as your cat knows about you.

> When I play with my cat, who knows whether she is not amusing herself with me more than I with her.
>
> MICHEL DE MONTAIGNE

> It doesn't matter if a cat is black or white, so long as it catches mice.
>
> DENG XIAOPING

Samuel Langhorne Clemens (Mark Twain) must surely have loved cats for he kept many and was a keen observer of their behavior. In his books and short stories, Mark Twain used cats to illustrate lessons.

> A man who carries a cat by the tail learns something he can learn in no other way.

> When a man loves cats, I am his friend and comrade, without further introduction.
>
> "An Incident," *Who Is Mark Twain?*

> The person that had took a bull by the tail once had learnt sixty or seventy times as much as a person that hadn't, and said a person that started in to carry a cat home by the tail was getting knowledge that was always going to be useful to him, and warn't ever going to grow dim or doubtful.
>
> MARK TWAIN, *Tom Sawyer Abroad*

You may say a cat uses good grammar. Well, a cat does—but you let a cat get excited once; you let a cat get to pulling fur with another cat on a shed, nights, and you'll hear grammar that will give you the lockjaw. Ignorant people think it's the noise which fighting cats make that is so aggravating, but it ain't so; it's the sickening grammar they use.

Mark Twain, *A Tramp Abroad*

A cat is more intelligent than people believe, and can be taught any crime.

Mark Twain, Notebook, 1895

Of all God's creatures there is only one that cannot be made the slave of the lash. That one is the cat. If man could be crossed with the cat it would improve man, but it would deteriorate the cat.

Mark Twain, Notebook, 1894

By what right has the dog come to be regarded as a "noble" animal? The more brutal and cruel and unjust you are to him the more your fawning and adoring slave he becomes; whereas, if you shamefully misuse a cat once she will always maintain a dignified reserve toward you afterward—you will never get her full confidence again.

Mark Twain, *A Biography*

I urged that kings were dangerous. He said, then have cats. He was sure that a royal family of cats would answer every purpose. They would be as useful as any other royal family, they would know as much, they would have the same virtues and the same treacheries, the same disposition to get up shindies with other royal cats, they would be laughably vain and absurd and never know it, they would be wholly inexpensive, finally, they would have as sound a divine right as any other royal house. . . . The worship of royalty being founded in unreason, these graceful and harmless cats would easily become as sacred as any other royalties, and indeed more so, because it would presently be noticed that they hanged nobody, beheaded nobody, imprisoned nobody, inflicted no cruelties or injustices of any sort, and so must be worthy of a deeper love and reverence than the customary human king, and would certainly get it.

Mark Twain, *A Connecticut Yankee in King Arthur's Court*

The cat sat down. Still looking at us in that disconcerting way, she tilted her head first to one side and then the other, inquiringly and cogitatively, the way a cat does when she has struck the unexpected and can't quite make out what she had better do about it. Next she washed one side of her face, making such an awkward and unscientific job of it that almost anybody would have seen that she was either out of practice or didn't know how. She stopped with the one side, and looked bored, and as if she had only been doing it to put in the time, and wished she could think of something else to do to put in some more time. She sat a while, blinking drowsily, then she hit an idea, and looked as if she wondered she hadn't thought of it earlier. She got up and went visiting around among the furniture and belongings, sniffing at each and every article, and elaborately examining it. If it was a chair, she examined it all around, then jumped up in it and sniffed all over its seat and its back; if it was any other thing she could examine all around, she examined it all around; if it was a chest and there was room for her between it and the wall, she crowded herself in behind there and gave it a thorough overhauling; if it was a tall thing, like a washstand, she would stand on her hind toes and stretch up as high as she could, and reach across and paw at the toilet things and try to rake them to where she could smell them; if it was the cupboard, she stood on her toes and reached up and pawed the knob; if it was the table she would squat, and measure the distance, and make a leap, and land in the wrong place, owing to newness to the business; and, part of her going too far and sliding over the edge, she would scramble, and claw at things desperately, and save herself and make good; then she would smell everything on the table, and archly and daintily paw everything around that was movable, and finally paw something off, and skip cheerfully down and paw it some more, throwing herself into the prettiest attitudes, rising on her hind feet and curving her front paws and flirting her head this way and that and glancing down cunningly at the object, then pouncing on it and spatting it half the length of the room, and chasing it up and spatting it again, and again, and racing after it and fetching it another smack—and so on and so on; and suddenly she would tire of it and try to find some way to get to the top of the cupboard or the wardrobe, and if she couldn't she would look troubled and disappointed; and toward the last, when you could see she was getting her bearings well lodged in her head and was satisfied with the place and the arrangements, she relaxed her intensities, and got to purring a little to herself, and praisefully waving her tail between inspections—and at last she was done—done, and everything satisfactory and to her taste.

Being fond of cats, and acquainted with their ways, if I had been a stranger and a person had told me that this cat had spent half an hour in that room before, but hadn't happened to think to examine it until now, I should have been able to say with conviction, "Keep an eye on her, that's no orthodox cat, she's an imitation, there's a flaw in her make-up, you'll find she's born out of wedlock or some other arrested-development accident has happened, she's no true Christian cat, if I know the signs."

Mark Twain, No. 44, *The Mysterious Stranger*

> That's the way with a cat, you know—any cat; they don't give a damn for discipline. And they can't help it, they're made so. But it ain't really insubordination, when you come to look at it right and fair—it's a word that don't apply to a cat. A cat ain't ever anybody's slave or serf or servant, and can't be—it ain't in him to be. And so, he don't have to obey anybody. He is the only creature in heaven or earth or anywhere that don't have to obey somebody or other, including the angels. It sets him above the whole ruck, it puts him in a class by himself. He is independent. You understand the size of it? He is the only independent person there is. In heaven or anywhere else. There's always somebody a king has to obey—a trollop, or a priest, or a ring, or a nation, or a deity or what not—but it ain't so with a cat. A cat ain't servant nor slave to anybody at all. He's got all the independence there is, in Heaven or anywhere else, there ain't any left over for anybody else. He's your friend, if you like, but that's the limit—equal terms, too, be you king or be you cobbler; you can't play any I'm-better-than-you on a cat—no, sir! Yes, he's your friend, if you like, but you got to treat him like a gentleman, there ain't any other terms. The minute you don't, he pulls freight.
>
> Mark Twain, *The Refuge of the Derelicts*

How are small cats incorporated into poetry?

Cats appear in all sorts of poetry, from the children's nursery rhyme of the "Three Little Kittens" to T. S. Eliot's *Old Possum's Book of Practical Cats*. Eliot's book contains some lines that those who study small cats might want to contemplate: "But above and beyond there's still one name left over / And that is the name that you never will guess / The name that no human research can discover." There is even a poem by M. L. Squier called "The Ocelot." William Wordsworth's "The Kitten and the Falling Leaves" typifies the joy that cats experience in nature.

> That way look, my infant, lo!
> What a pretty baby-show!
> See the kitten on the wall,
> sporting with the leaves that fall.
> Withered leaves - one - two and three
> from the lofty elder tree.
> Though the calm and frosty air,
> of this morning bright and fair.
> Eddying round and round they sink,
> softly, slowly; one might think.
> From the motions that are made,

every little leaf conveyed
Sylph or Faery hither tending,
to this lower world descending.
Each invisible and mute,
in his wavering parachute.

But the Kitten, how she starts,
crouches, stretches, paws, and darts!
First at one, and then its fellow,
just as light and just as yellow.
There are many now - now one,
now they stop and there are none:
What intenseness of desire,
in her upward eye of fire!
With a tiger-leap half-way,
now she meets the coming prey.
lets it go as fast, and then;
Has it in her power again.
Now she works with three or four,
like an Indian conjuror;
quick as he in feats of art,
far beyond in joy of heart.
Where her antics played in the eye,
of a thousand standers-by,
clapping hands with shout and stare,
what would little Tabby care!
For the plaudits of the crowd?
Over happy to be proud,
over wealthy in the treasure
of her exceeding pleasure!

How are small cats incorporated into literature?

While visiting the McGill University library in Montreal, Canada, JGS had the unexpected privilege of seeing quite literally how often small cats have been incorporated into poetry and literature. A kindly librarian, knowing his interest in small cats, led him to several long shelves lined front-to-back with thousands of books that had something to do with cats.

Clearly, it is impossible to describe the enormous variety of ways small cats have been incorporated into poetry and literature. However, in our opinion there are two very special small cats that are quintessential characters in classic literature: Charles Lutwidge Dodgson's (Lewis Carroll)

Cheshire cat in *Alice's Adventures in Wonderland* (1865) and Edgar Allan Poe's "The Black Cat" (1843).

The Cheshire Cat, Lewis Carroll. The Cheshire cat is famous for its huge grin, and its ability to appear in different forms such as a head without a body, and a grim without either head or body, and for its engaging yet confounding conversations with Alice. After tumbling into a rabbit burrow, Alice repeatedly encounters the Cheshire cat. A memorable experience comes when Alice sees a grin floating in the air. Slowly Alice recognizes the grin to be that of the Cheshire cat. The cat is sitting in a tree.

Later the Queen of Hearts condemns the Cheshire cat to death by beheading. The Cheshire cat transforms into a head and characteristic grim without a body. Later the body and head of the Cheshire cat slowly fades, but its grin remains. At one point in the story, the Cheshire cat asks Alice what happened to the baby she was minding. Alice tells the Cheshire cat that the baby turned into a pig and ran away.

The Cheshire cat, along with many other characters in *Alice's Adventures in Wonderland*, has become an icon in our popular culture. Indeed, long after the author's real name (not Lewis Carroll), and actual title of the book (not *Alice in Wonderland*) are forgotten, and the body and head of the Cheshire cat disappear, its grin will still be with us. At San Francisco's Exploratorium there is an exhibit called The Cheshire Cat that illustrates precisely this phenomenon.

The Black Cat, Edgar Allan Poe. Just as the title suggests, *The Black Cat* is a dark tale indeed. After a long and loving relationship with his pet black cat Pluto, the narrator of the story becomes an alcoholic, begins to hate Pluto, gouges out its eye with a knife, and eventually hangs the cat to death in his backyard. That very night, his house burns to the ground except for one wall. On the wall is the image of Pluto hanging by his neck.

Later, the narrator finds a black cat nearly identical in appearance to Pluto (including having just one eye) at a local tavern. Perhaps, this cat is an incarnation of Pluto, another of Pluto's nines lives! Once again, the narrator's loving relationship turns to hate. When he attempts to kill the cat with an axe, his wife stops him. In his rage, he turns on his wife, killing her with the axe. He then hides the body of his deceased wife within a brick wall in his basement. Pluto's incarnation is, of course, missing.

Here are two other odes to cats in literature:

> He will kill mice, and he will be kind to babies when he is in the house, just as long as they do not pull his tail too hard. But when he has done that, and between times, and when the moon gets up and night comes, he is the Cat

that walks by himself, and all places are alike to him. Then he goes out to the Wet Wild Woods or up the Wet Wild Trees or on the Wet Wild Roofs, waving his wild tail and walking by his wild lone.

Rudyard Kipling, "The Cat That Walked by Himself"

When God made the world, He chose to put animals in it, and decided to give each whatever it wanted. All the animals formed a long line before His throne, and the cat quietly went to the end of the line. To the elephant and the bear He gave strength, to the rabbit and the deer, swiftness; to the owl, the ability to see at night, to the birds and the butterflies, great beauty; to the fox, cunning; to the monkey, intelligence; to the dog, loyalty; to the lion, courage; to the otter, playfulness. And all these were things the animals begged of God. At last he came to the end of the line, and there sat the little cat, waiting patiently. "What will YOU have?" God asked the cat.
The cat shrugged modestly. "Oh, whatever scraps you have left over. I don't mind."

"But I'm God. I have everything left over."

"Then I'll have a little of everything, please."

And God gave a great shout of laughter at the cleverness of this small animal, and gave the cat everything she asked for, adding grace and elegance and, only for her, a gentle purr that would always attract humans and assure her a warm and comfortable home.

But he took away her false modesty.

Lenore Fleischer, "When God Made Cats"

Chapter 12

"Small Catology"

Which species are best known?

It's important to appreciate the differences between a specialist's understanding of small cats and the general public's understanding of small cats. Even specialists find small cat identification confounding. These differences highlight the difficulties faced by conservation biologists in the global battle to save small cat species from extinction. First, let's consider from a global perspective the general public's understanding of small cats. There is little doubt that awareness of many species has grown as a result of wildlife documentaries presented on televisions all over the world. However, not all wild cats have been the stars of their own shows.

It's safe to say that the big cats have preferentially greatly benefitted from having been featured in wildlife documentaries and housed in public zoos all over the world. It's also true that most big cats are just plain easy to recognize, and out in the wild can't hide from the camera crew. Must the lion, the King of Beasts, hide? Certainly not. Who among the human residents on earth would not recognize a tiger? Even people living on remote islands in the middle of the Pacific Ocean know a tiger when they see one on television or in a zoo.

To a lesser extent the same is true for the remaining five of the word's biggest wild cats: cheetah, leopard, jaguar, snow leopard, and puma. For example, while working near a remote village in Nepal, JGS was surprised to hear locals in a small café discussing the natural history of the jaguar (that occurs in the Americas). Such awareness is made possible by the mass media and an insatiable desire by people all over the world to learn more about wildlife. The impact has been nothing less than profound. However, the absence of such awareness can be debilitating for conservation efforts. It's safe to say these Nepalese villagers knew far more about jaguars than they did about the Asiatic golden cat living in the foothills near their village.

Whereas big cats are nearly all widely recognized by the general public, not a single species of small cat is nearly as widely known. Instead, some small cats are generally only known regionally if they are known at all. Thus, it's important to appreciate that small cats remain largely unknown beyond the places where they live, and some species are not even known or recognized locally. But the situation is even worse than this. Sadly, even people working in conservation in their own country often confuse one species of small cat for another. JGS is grateful when people ask him to examine a picture, most often taken by a remotely triggered camera, and identify which species of small cat they photographed. In this way, mistaken identifications that are the bane of conservation efforts can be avoided.

Consider a small cat found in North America. The Canada lynx, for instance, would be recognized by most people living in Canada, where the cat is widely distributed. In the United States the best-known small cats are the bobcat and Canada lynx. The ocelot is widely recognized in Central and northern South America—its geographic distribution. However, the ocelot is not found in the high Andes of Argentina, Bolivia, Chile, and Peru and would therefore not be known to the people that live there. Instead, the people of the altiplano recognize and are able to tell the difference between the Andean cat and the pampas cat, something no one else in the entire world, aside from a small cat authority, could do.

In parts of Europe the wildcat and the Eurasian lynx are recognized, and in Spain there is little doubt the wildcat and the Iberian lynx are best known, although the wildcat is easily confused with the domestic cat. In parts of Africa where the serval and caracal co-occur, these small cats would be recognized to be two distinct species, but outside Africa, few people would know the difference. In Southeast Asia, the leopard cat is the most widespread small cat and is therefore probably the most widely recognized of the small cats. However, conservationists working in Sumatra often confuse the leopard cat for the fishing cat (where the fishing cat was thought to occur but in fact does not).

In Mongolia, pastoralists easily recognize Pallas' cat since it naturally occurs there. However, where Pallas' cat does not occur, very few people beyond experts would recognize this beautiful small cat. In all of North, Central, and South America, Africa, Europe, and Australia, there is little doubt the general public would mistake Pallas' cat for a domestic cat. Such lack of awareness leaves conservation efforts desperate for critical funding.

You might agree that small cats suffer from a public relations problem.

Certainly there is general agreement that the big cats are far more widely recognized than the small cats. However, this is generally true of all wildlife. For instance, the best-known species of primates are the chimpanzee and gorillas, the largest of the primates. Who would not recognize an elephant? But how many people would recognize a guigna?

Several species of small cats have received the attention of students and scientists, while most remain unstudied; the details of their lives remain an enigma. In North America, the bobcat and the Canada lynx have received a great deal of attention. The ecology and biology of both species is well known and has received the attention of countless state and provincial wildlife departments. In Central and South America the ocelot has received much attention, and because it is considered endangered, the Andean cat has received much attention recently. The diminutive guigna of Chile in South America has also been the subject of several recent studies. In contrast, the bay cat, endemic to Borneo, remains as unknown as it was when it was discovered in the mid-1800s.

The European small cats, the wildcat, Eurasian lynx, and Iberian lynx, have all received much attention. In Africa, the small black-footed cat has been the beneficiary of ongoing long-term studies by Alex Sliwa. The wildcat, serval, and caracal have also received some attention in Africa. In Asia, the most common small cat is the leopard cat. Hence, the leopard cat is the best-studied small cat species in Asia. But not a single study has been made of the flat-headed cat or bay cat. Pallas' cat has recently received attention in Russia and Mongolia.

Members of the lynx lineage—bobcat, Iberian lynx, Eurasian lynx, and Canada lynx—share a unique distinction. This is the only lineage where all members have been the subject of scientific studies and reintroduction programs. Despite a lack of knowledge of the bobcat in Mexico, the lynx lineage is the best-known lineage of small cats.

Which species are least known?

From the general public's perspective, small cats are best known regionally. Outside the region where small cats naturally occur, knowledge drops precipitously. From a global vantage point, no small cats are nearly as well known as any of the big cats.

The least known small cat is the endangered bay cat that is only found on the island of Borneo. Our ignorance does not stem from a lack of looking for the cat. Authorities insist the bay cat is the rarest cat on earth. We simple don't know enough about the cat to say one way or the other just how rare it really is. We believe that bay cat has always been rare. The great naturalist and contemporary of Charles Darwin, Alfred Russel Wallace,

Bobcats are the most widely recognized small cats in North America. Photo © Ken Adelman

In Europe, the Eurasian lynx might be the most widely recognized small cat. Photo © Neville Buck

collected the first bay cat in 1858. Because the specimen Wallace had collected in Kuching, Sarawak, and sent back to London was in such bad condition, it was decided to wait for a second specimen before announcing to the world that a small cat new to science had been discovered. Wallace never found a second bay cat. In 1864, based upon Wallace's single militated skin and partial skull, the announcement that a small cat new to science had been discovered on Borneo was made.

In 1883, Daniel Giraud Elliot wrote that *nothing* was known of the bay cat in the wild. Today we can write: *little* is known of the bay cat in the wild. The bay cat was the last species of wildcat to be discovered. Other small cats have subsequently been named but are now recognized to be subspecies of already named species, or molecular analysis has revealed that what

As its name indicates, the fishing cat eats, among other prey, fish, and does not hesitate to enter water and submerse its head in search of its prey. Photo © Nancy Verdermey

was once thought to be a single species should in fact be considered two species.

Without question the most poorly known species occur in Asia, where conservation issues are most acute. Along with the bay cat, the flat-headed cat and the marbled cat are very poorly understood and have not received the attention they deserve. The African golden cat, which occurs in the rainforest belt that extends across central Africa, is also very poorly understood.

More information is needed about the habits and habitat needs of two contrasting specialists: the fishing cat and sand cat. Until recently, the fishing cat was believed to occur in Sumatra. The fishing cat was once found in Java, however, the last individual fishing cat to be seen was shot by a hunter in 1932. Has the fishing cat gone extinct in Java? We do not know.

The sand cat inhabits the deserts of North Africa and the Middle East. Beyond this general description, not much is known. Conservationists believe sand cats have disappeared from many areas within its geographic range, but hard evidence is lacking, and no one has made the sand cat a top priority.

Under the umbrella of the jaguar, which is the subject of more than a dozen conservation efforts, in southern Central America and northern South America more information is dearly needed regarding the status of the tigrina, also known as the oncilla or small spotted cat. The Andean cat, the only member of the Felidae in the Americas considered endangered, must become the subject of a long-term monitoring program. Several populations of the more widespread pampas cat seem to have disappeared, but critical information is missing.

As a group, small cats have not been well studied and the conservation status of many species is unknown. In recent years, more efforts have been made, but a lack of global awareness regarding the plight of the world's small cats is hampering conservation efforts.

How do scientists recognize individual small cats?

One of the most frequently asked questions about small cats is: how many are there?

Determining just how many individuals of a certain species there are is one of the most challenging problems in small cat conservation. As might be expected, it's not possible to simply drive around and count them.

First, the population can only be estimated. The population will never be known precisely unless there are so few, as with the Iberian lynx, that we know where each individual lives, and we can count them. To estimate the population requires that individuals be known, or at least individuals can be differentiated. For most small cat species, this is impossible because, for those species without spots or stripes, it is impossible to tell individuals apart.

Census bureaus all over the world are responsible for determining how many people live in a certain place. The same is true for conservation biologists who study small cats. We want to know how many individuals of a certain species are found in a known area. The proper identification of individuals is necessary to determine the population.

Individual identification assumes that each individual, like a human individual, has unique characters that can be easily seen in the field. Spot patterns are the most obvious trait that allows individuals to be identified. Without spot patterns, forget it! It turns out that spot patterns are unique to individuals and surprisingly to each side of each individual. That is, no two spotted individuals are alike and, moreover, the two sides of the same spotted individual are unique. Thus, both sides can serve to uniquely identify individual small spotted cats. However, this requires that both sides of the same individual must be seen *at the same time*, or at least not too long apart.

Brightly and distinctly spotted small cats such as the ocelot, margay, tigrina, leopard cat, marbled cat, clouded leopard, and Sunda clouded leopard are more easily distinguished. For small cats without unique and distinct identifying characters or coat patterns, differentiating individuals is difficult, if not, impossible. Some species like the Andean cat, guigna, pampas cat, Geoffroy's cat, Pallas' cat, sand cat, jaguarundi, Asiatic golden cat, African golden cat, caracal, bay cat, flat-headed cat, and fishing cat have few, if any, distinct markings, making individual identification impossible in the absence of other information.

Recent advances in molecular technology have enabled individual recognition of every species of small cat provided samples containing an individual's DNA can be obtained. DNA has been obtained from blood, hair, and scat samples, for instance. To obtain hair samples, hair snares can be used. Hair snares are posts wrapped in velcro and armed with an attractant—a scent lure—that attracts small cats. The idea is that the individual will rub against the scent post and leave a few hairs behind. Scat samples also yield DNA. To find scat, trained dogs, called scat-dogs, have been used. Scat-dogs are highly trained dogs that sniff out small cat scat. Laboratory analysis of the scat can identify the species and the individual that left it.

Appendix

Small Cats of the World

Order Carnivora
Family Felidae

Lineage	Scientific name	Common name	Geographic location
bay cat	*Catopuma temminckii*	Asiatic golden cat	Southeast Asia
	Catopuma badia	bay cat	Southeast Asia
	Pardofelis marmorata	marbled cat	Southeast Asia
caracal	*Caracal aurata*	African golden cat	Central Africa
	Caracal caracal	caracal	Africa, Middle East
	Leptailurus serval	serval	Africa
Felis	*Felis nigripes*	black-footed cat	Southern Africa
	Felis catus	domestic cat	Global
	Felis chaus	jungle cat	Nile Delta, Middle East
	Felis margarita	sand cat	North Africa, Middle East
	Felis silvestris	wildcat	Europe, Africa, Asia
leopard cat	*Prionailurus viverrinus*	fishing cat	Southeast Asia
	Prionailurus planiceps	flat-headed cat	Southeast Asia
	Prionailurus bengalensis	leopard cat	Asia
	Otocolobus manul	Pallas' cat	Central Asia
	Prionailurus rubiginosus	rusty-spotted cat	India, Sri Lanka
lynx	*Lynx rufus*	bobcat	North America
	Lynx Canadensis	Canada lynx	North America
	Lynx lynx	Eurasian lynx	Eurasia
	Lynx pardinus	Iberian lynx	Spain (Iberian peninsula)
ocelot	*Leopardus jacobita*	Andean cat	South America
	Leopardus geoffroyi	Geoffroy's cat	South America
	Leopardus guigna	guigna	South America
	Leopardus wiedii	margay	Central and South America
	Leopardus pardalis	ocelot	North, Central, and South America
	Leopardus colocolo	pampas cat	South America
	Leopardus tigrinus	tigrina	Central and South America
Panthera	*Neofelis nebulosa*	clouded leopard	Southeast Asia and Southern China
	Neofelis diardi	Sunda clouded leopard	Borneo, Sumatra
puma	*Puma yagouaroundi*	jaguarundi	Central and South America

Bibliography

Abreu, K. C., P. F. Moro-Rios, J. E. Silva-Pereira, J. M. D. Miranda, E. F. Jablonski, and F. C. Passos. 2008. Feeding habits of ocelot *(Leopardus pardalis)* in southern Brazil. *Mammalian Biology* 73:407–411.

Alda, F., J. Inoges, L. Alcaraz, J. Oria, A. Aranda, and I. Doadrio. 2008. Looking for the Iberian lynx in central Spain: A needle in a haystack? *Animal Conservation* 11:297–305.

Alderton, D. 1993. *Wild Cats of the World*. Blandford, UK.

Al-Jumaily, M. M. 1998. Review of the mammals of the Republic of Yemen. *Fauna of Arabia* 17:477–499.

Allayarov, A. M. 1964. Data on the ecology and geographical distribution of the jungle cat in Uzbekistan. *Uzbek Biological Journal*: 1–7.

Antón, M., and A. Turner. 1993. *The Big Cats and Their Fossil Relatives*. Columbia University Press, New York.

Aripse, R., D. Rumiz, and A. J. Noss. 2007. Six species of cats registered by camera trap surveys of tropical dry forest in Bolivia. *Cat News* 47:36–38.

Austin, S. C., M. E. Tewes, L. I. Grassman Jr., and N. J. Silvy. 2007. Ecology and conservation of the leopard cat *(Prionailurus bengalensis)* and clouded leopard *(Neofelis nebulosa)* in Khao Yai National Park, Thailand. *Acta Zoologica Sinica* 53:1–14.

Azlan, M., D. Noorafizah, and J. Sanderson. 2007. Historical records of felid collections in the Sarawak Museum. *Cat News* 47:8–9.

Baker, M. A., K. Nassar, L. Rifai, M. Qarqaz, W. Al-Melhim, and Z. Amr. 2003. On the current status and distribution of the jungle cat, *Felis chaus*, in Jordan (Mammalia: Carnivora). *Zoology in the Middle East* 30:5–10.

Belousova, A. V. 1993. Small Felidae of Eastern Europe, Central Asia and Far East: Survey of the state of populations. *Lutreola* 2:16–21.

Bergman, C. 1998. The almost-missing lynx: Spain's Donana National Park shelters one of the world's most endangered cats. *Natural History* 10:37–45.

Bezuijen, M. R. 2000. The occurrence of the flat-headed cat Prionailurus planiceps in south-east Sumatra. *Oryx* 34:222–226.

Bowland, J. 1990. Servals: Wetland cats. *Endangered Wildlife* 1:4–5.

Boy, G. 2003. Phantom feline. *Swara* 26:25–43.

Breitenmoser, U., C. Breitenmoser-Würsten, S. Capt, A. Molinari-Jobin, P. Molinari, P. and F. Zimmermann. 2007. Conservation of the lynx *(Lynx lynx)* in the Swiss Jura Mountains. *Wildlife Biology* 13:340–355.

Buckley-Beason, V. A., W. E. Johnson, W. G. Nash, et al. 2006. Molecular evidence for species-level distinctions in clouded leopards. *Current Biology* 16:2371–2376.

Carroll, C. 2005. Following the stealth hunter. *National Geographic*, November: 66–77.

Carroll, L. 1865. *Alice's Adventures in Wonderland*. Macmillan and Co., London.

Chakraborty, S. 1978. The rusty-spotted cat, *Felis rubiginosa* I. Geoffroy, in Jammu and Kashmir. *Journal of the Bombay Natural History Society* 75:478–479.

Childs, J. L., E. B. McCain, A. M. Childs, and J. Brun. 2007. The borderlands jaguar detection project: A report on the jaguar in southeastern Arizona. *Wild Cat News* 1–9.

Clutton-Brock, J. 2000. *The British Museum Book of Cats*. The British Museum Press, London.

Cunha Serra, R., and P. Sarmento. 2007. The Iberian lynx in Portugal: Conservation status and perspectives. *Cat News* 45:15–16.

Dale-Green, P. 1963. *The Cult of the Cat*. Weathervane Books, New York.

Datta, A., M. O. Anand, and R. Naniwadekar. 2008. Empty forests: Large carnivore and prey abundance in Namdapha National Park, north-east India. *Biological Conservation* 141:1429–1435.

Delgado, E., L. Villalba, J. Sanderson, C. Napolitano, M. Berna, and J. Esquivel. 2004. Capture of an Andean cat in Bolivia. *Cat News* 40:2.

de Oliveira, T. G. 1993. *Neotropical Cats: Ecology and Conservation*. EDFUMA, São Paulo.

Dillon, A., and M. J. Kelly. 2008. Ocelot home range, overlap and density: Comparing radio telemetry with camera trapping. *Journal of Zoology* 275:391–398.

Dookia, S. 2007. Sighting of Asiatic wildcat in Gogelao enclosue, Nagaur in Thar Desert of Rajasthan. *Cat News* 46:17–18.

Downey, P. 2005. The margay of El Cielo. *Wild Cat News* 1–2.

Dragesco-Joffe, A. 1993. The sand cat: A formidable snake hunter. *La Vie Sauvage du Sahara*, 129–133. Delachaux et Niestlé, Lausanne, France.

Driscoll, C. A., M. Menotti-Raymond, A. L. Roca, et al. 2007. The Near Eastern origin of cat domestication. *Sciencexpress* 1–6, 28 June 2007, 10.1126/science .1139518.

Dubey, Y. 1999. Sighting of rusty spotted cat *Prionailurus rubiginosus* in Tadoba Andhari Tiger Reserve, Maharashtra. *Journal of the Bombay Natural History Society* 96:310–311.

Duckworth, J. W., C. M. Poole, R. J. Tizard, J. L. Walston, and R. J. Timmins. 2005. The jungle cat *Felis chaus* in Indochina: A threatened population of a widespread and adaptable species. *Biodiversity and Conservation* 14:1263–1280.

Duckworth, J. W., C. R. Sheppard, G. Semiadi, et al. 2010. Does the fishing cat inhabit Sumatra? *Cat News* 51:4–9.

Elliot, D. G. 1878–1883. *A Monograph of the Felidae*. Published by the author.

Emmons, L. H., P. Sherman, D. Bolster, A. Goldizen, and J. Terborgh. 1989. Ocelot behavior in moonlight. *Advances in Neotropical Mammalogy*: 233–242.

Ewer, R. E. 1973. *The Carnivores*. Cornell University Press, Ithaca, New York.

Fox, J. L., and T. Dorji. 2007. High elevation record for occurrence of manul or Pallas cat on the northWestern Tibetan plateau, China. *Cat News* 46:35.

Geertsema, A. 1991. The servals of Gorigor. *Natural History:* 52–61.

Germain, E, S. Benhamou, and M.-L. Poule. 2008. Spatio-temporal sharing between the European wildcat, the domestic cat and their hybrids. *Journal of Zoology* 276:195–203.

Gittleman, J. L., ed. 1989. *Carnivore Behavior, Ecology, and Evolution*. Cornell University Press, Ithaca, New York.

Glick, D. 2006. Of lynx and men. *National Geographic*, January: 56–67.

Grassman L. I. Jr., A. M. Haines, J. E. Janecka, and M. E. Tewes. 2006. Activity periods of photo-captured mammals in north-central Thailand. *Mammalia:* 306–309.

Grassman L. I. Jr., K. Kreetiyutanont, and M. E. Tewes. 2002. Survey and status of the carnivore community in northeastern Thailand. *Tiger Paper* 29:1–3.

Grassman, L. I. Jr., and M. E. Tewes. 2002. Marbled cat pair in northeastern Thailand. *Cat News* 36:19.

Grassman, L. I. Jr., and M. E. Tewes. 2003. Jaguarundi: The weasel cat of Texas. *South Texas Wildlife* 8:1–2.

Grigione, M. M., K. Menke, C. A. López-Gonzólez, R. List, A. Banda, J. Carrera, R. Carrera, A. J. Giordano, J. Morrison, M. Sternberg, R. Thomas, and B. Van Pelt. 2009. Identifying potential conservation areas for felids in the USA and Mexico: Integrating reliable knowledge across an international border. *Oryx* 43:78–86.

Gusset, M., M. J. Swarner, L. Mponwanek, K. Keletiele, and J. W. McNutt. 2009. Human-wildlife conflict in northern Botswana: Livestock predation by endangered African wild dog *(Lycaon pictus)* and other carnivores. *Oryx* 43: 67–72.

Harveson, P. M., M. E. Tewes, G. L. Anderson, and L. L. Laack. 2004. Habitat use by ocelot in south Texas: Implications for restoration. *Wildlife Society Bulletin* 32:948–954.

Heilbrun, R. D., N. J. Silvy, M. J. Peterson, and M. E. Tewes. 2006. Estimating bobcat abundance using automatically triggered cameras. *Wildlife Society Bulletin* 34 (1): 69–73.

Hetherington, D. 2008. The history of the Eurasian lynx in Britain and the potential for its reintroduction. *British Wildlife:* 77–86.

Hoffmann, T. W. 1975. Wild cats of Sri Lanka. *Loris* 13:286.

Hunter, L. 2000. The serval: High-rise hunter. *Africa—Environment & Wildlife*, July: 34–40.

Hunter, L. 2003. Africa's cryptic cats. *Africa Geographic*, August: 43–49.

Iriarte, A., and J. G. Sanderson. 1999. Home-range and activity patterns of kodkod Oncifelis guigna on Isla Grande de Chiloé, Chile. *Cat News* 30:27.

Jackson, P., A. F. Jackson, R. Dallet, J. de Crem. 1996. *Les félins*. Delahaux et Niestlé, Lausanne.

Janis, C. 1994. The sabertooth's repeat performances. *Natural History* 4:78–83.

Johnson, M. A., P. M. Saraiva, and D. Coelho. 1999. The role of gallery forests in the distribution of Cerrado mammals. *Rev. Brasil. Biol.* 59:421–427.

Johnson, W. E., E. Eizirik, J. Pecon-Slattery, W. J. Murphy, A. Antunes, E. Teeling, and S. J. O'Brien. 2006. The late Miocene radiation of modern felidae: A genetic assessment. *Science* 311:73–77.

Kawanishi, K., and M. E. Sunquist. 2008. Food habits and activity patterns of the Asiatic golden cat *(Catopuma temminckii)* and dhole *(Cuon alpinus)* in primary rainforest of peninsular Malaysia. *Mammal Study* 33:173–177.

Khan, A. A., and M. A. Beg. 1986. Food of some mammalian predators in the cultivated areas of Punjab. *Pakistan Journal of Zoology* 18:71–79.

Kitchener, A. 1991. *The Natural History of Wild Cats.* Cornell University Press, Ithaca, New York.

Kitchener, A. C., M. A. Beaumont, D. Richardson. 2006. Geographical variation in the clouded leopard, *Neofelis nebulosa,* reveals two species. *Current Biology* 16:2377–2383.

Koehler, G. M., B. T. Maletzke, J. A. von Kienast, K. B. Aubry, R. B. Wielgus, and R. H. Naney. 2008. Habitat fragmentation and the persistence of lynx populations in Washington state. *Journal of Wildlife Management* 72:1518–1524.

Larrucea, E. S., G. Serra, M. M. Jaeger, R. H. Barrett. 2007. Censusing bobcats using remote cameras. *Western North American Naturalist* 67:538–548.

Leyhausen, P. 1979. *Cat Behavior.* Garland STPM Press, New York.

Liat, L. B., and I. A. Rahman bin Omar. 1961. Observations on the habits in captivity of two species of wild cats, the leopard cat and the flat-headed cat. *Malayan Natural History Journal* 15:48–51.

Lucherini, M., E. Lugengos Vidal, and M. J. Merino. 2008. How rare is the rare Andean cat? *Mammalia* 72:95–101.

Lucherini, M., and M. J. Merino. 2008. Perceptions of human-carnivore conflicts in the High Andes of Argentina. *Mountain Research and Development* 28:81–85.

Lynch, G. S., J. D. Kirby, R. J. Warren, and L. M. Conner. 2008. Bobcat spatial distribution and habitat use relative to population reduction. *Journal of Wildlife Management* 72:107–112.

Lyra-Jorge, M. C., G. Ciocheti, and V. R. Pivello. 2007. Carnivore mammals in a fragmented landscape in northeast of São Paulo State, Brazil. *Biodiversity and Conservation* 17:1573–1580.

Malek, J. 1993. *The Cat in Ancient Egypt.* The British Museum Press, London.

Mallon, D. P. 1985. The mammals of the Mongolian People's Republic. *Mammal Review* 15 (2): 71–102.

Manfredi, C., M. Lucherini, A. D. Canepuccia, and E. B. Casanave. 2004. Geographical variation in the diet of Geoffroy's cat *(Oncifelis geoffroyi)* in pampas grassland of Argentina. *Journal of Mammalogy* 85:1111–1115.

Manfredi, C., L. Soler, M. Lucherini, E. B. Casanave, K. A. Abernethy. 2006. Home range and habitat use by Geoffroy's cat *(Oncifelis geoffroyi)* in a wet grassland in Argentina. *Journal of Zoology:* 1–7.

Mayr, E. 1963. *Animal Species and Evolution.* Harvard University Press, Cambridge, MA.

Miller, S. D., and D. D. Everett, eds. 1986. *Cats of the World: Biology, Conservation, and Management.* National Wildlife Foundation, Washington, DC.

Mivart, St. G. 1892. *The Cat.* Charles Scribner's Sons, New York.

Mohd-Azlan, J. 2009. Celebrity animal: Borneo's Houdini. *Malaysian Naturalist* 63 (1): 10–11. Photographs by J. G. Sanderson.

Mohd-Azlan, J., and J. Sanderson. 2007. Geographic distribution and conservation status of the bay cat *Catopuma badia,* a Bornean endemic. *Oryx* 41 (3): 1–4.

Mohd-Azlan, J., and D. S. K. Sharma. 2006. The diversity and activity patterns of wild felids in a secondary forest in peninsular Malaysia. *Oryx* 40:36–41.

Molteno, A. J., A. Sliwa, and P. R. K. Richardson. 1998. The role of scent marking in a free-ranging, female black-footed cat *(Felis nigripes)*. *Journal of the Zoological Society of London* 245:35–41.

Morris, D. 1969. *The Human Zoo*. McGraw-Hill, New York.

Morris, D. 1993. *Catwatching*. Random House, New York.

Mukherjee, S., and C. Groves. 2007. Geographic variation in jungle cat *(Felis chaus* Schreber, 1777) (Mammalia, Carnivora, Felidae) body size: Is competition responsible? *Biological Journal of the Linnean Society* 92:163–172.

Munkhtsog, B., S. Ross, and M. Brown. 2004. Home range characteristics and conservation of Pallas' cat in Mongolia. Unpublished manuscript.

Murray, D. L., T. D. Steury, and J. D. Roth. 2008. Assessment of Canada lynx research and conservation needs in the southern range: Another kick at the cat. *Journal of Wildlife Management* 72:1463–1472.

Napolitano, C., M. Bennett, W. E. Johnson, S. J. O'Brien, P. A. Marquet, I. Barria, E. Poulin, and A. Iriarte. 2008. Ecological and biogeographical inferences on two sympatric and enigmatic Andean cat species using genetic identification of faecal samples. *Molecular Ecology* 17:678–690.

Nayerul Haque, M. D., and V. S. Vijayan. 1993. Food habits of the fishing cat in Keoladeo National Park, Bharatpur, Rajasthan. *Journal of the Bombay Natural History Society* 90:498–500.

Nowell, K., and P. Jackson. 1996. *Wild Cats: Status Survey and Conservation Action Plan*. Burlington Press, Cambridge.

O'Brien, J., S. Devillard, L. Say, H. Vanthomme, F. Léger, S. Ruette, and D. Pontier. 2009. Preserving genetic integrity in a hybridising world: Are European wildcats *(Felis silvestris silvestris)* in eastern France distinct from sympatric feral domestic cats? *Biodiversity and Conservation* 18: 2351–2360.

Oliveira, R., R. Godinho, E. Randi, and P. C. Alves. 2008. Hybridization versus conservation: Are domestic cats threatening the genetic integrity of wildcats *(Felis silvestris silvestris)* in Iberian peninsula? *Phil. Trans. R. Soc. B* 363 (1505): 2953–2961.

Olbricht, G., and A. Sliwa. 1995a. Analyse der Jugendentwicklung von Schwarzfußkatzen *(Felis nigripes)* im Zoologischen Garten Wuppertal im Vergleich zur Literatur. *Zoologischer Garten* 65 (4): 224–236.

Olbricht, G., and A. Sliwa. 1995b. Comparative development of juvenile black-footed cats at Wuppertal Zoo and elsewhere. *International Studbook for the Black-footed cat* (Felis nigripes), 8–20. Zooligischer Garten der Stadt Wuppertal, Germany.

Olbricht, G., and A. Sliwa. 1997. In situ and ex situ observations and management of black-footed cats *Felis nigripes*. *International Zoological Yearbook* 35:81–89.

Pereira, J. A., N. G. Fracassi, and M. M. Uhart. 2006. Numerical and spatial responses of Geoffroy's cat *(Oncifelis geoffroyi)* to prey decline in Argentina. *Journal of Mammalogy* 87:1132–1139.

Poe, E. A. 1843. "The Black Cat." *The Saturday Evening Post*, 19 August, New York.

Rajaratnam, R., M. Sunquist, L. Rajaratnam, and L. Ambu. 2007. Diet and habitat selection of the leopard cat *(Prionailurus bengalensis borneoensis)* in an agricultural landscape in Sabah, Malaysian Borneo. *Journal of Tropical Ecology* 23:209–217.

Randi, R. 2008. Detecting hybridization between wild species and their domesticated relatives. *Molecular Ecology* 17:285–293.

Rao, M., T. Myint, T. Zaw, and S. Htun. 2005. Hunting patterns in tropical forests adjoining the Hkakaborazi National Park, north Myanmar. *Oryx* 39:292–300.

Ray, J. C., and M. E. Sunquist. 2001. Trophic relations in a community of African rainforest carnivores. *Oecologia* 127:395–408.

Rice, J. 1997. Sand trapped. *Natural History*, July–August: 78–79.

Ruiz, A. 2001. *The Spirit of Ancient Egypt*. Algora Publishing, New York.

Sánchez-Cordero, V., D. Stockwell, S. Sarkar, H. Liu, C. R. Stephens, and J. Giménez. 2008. Competitive interactions between felid species may limit the southern distribution of bobcats *(Lynx rufus)*. *Ecography* 31:757–764.

Sanderson, J. G. 1997. Oncifelis guigna, the *guigna*. *Cat News* 26:16.

Sanderson, J. G. 1999. Andean mountain cats *(Oreailurus jacobita)* in northern Chile. *Cat News* 30:25–26.

Sanderson, J. G. 2007. No mean cat feat. *Science* 310:1151.

Sanderson, J. G. 2009. How the fishing cat came to occur in Sumatra. *Cat News* 50:6–9.

Sanderson, J. G., and M. E. Sunquist. 1998. Ecology and behavior of the kodkod in a highly-fragmented, human-dominated landscape. *Cat News* 28:16–17.

Sanderson, J., M. E. Sunquist, and A. Iriarte. 2002. Natural history and landscape use of guignas *(Oncifelis guigna)* on Isla Grande de Chiloé, Chile. *Journal of Mammalogy* 83 (2): 608–613.

Sanderson, J. G., and M. Trolle. 2005. Monitoring elusive mammals. *American Scientist* 93:148–155.

Sanderson, J. G., and L. Villalba. 2006. Sacred cat of the Andes. *Wild Cat News* 2 (2): 7–11.

Sarmento, P., J. Cruz, C. Eira, and C. Fonseca. 2009. Spatial colonization by feral domestic cats *Felis catus* of former wildcat *Felis silvestris silvestris* home ranges. *Acta Theriologica* 54:31–38.

Schmidt, K., W. Jedrzejewski, H. Okarma, and R. Kowalczyk. 2009. Spatial interactions between grey wolves and Eurasian lynx in Bialowieza Primeval Forest, Poland. *Ecological Research* 24:207–214.

Schmidt, K., R. Kowalczyk, J. Ozolins, P. Männil, and J. Fickel. 2009. Genetic structure of the Eurasian lynx population in north-eastern Poland and the Baltic states. *Conservation Genetics* 10:497–501.

Scrocchi, G. J., and S. P. Halloy. 1986. Notas sistemáticas, ecológicas, etológicas y biogeográficas sobre el gato Andino, *Felis jacobita* Cornalia (Felidae, Carnívora). *Acta Zoológica Lilloana* 38 (2): 157–170.

Seidensticker, J., and S. Lumpkin. 2004. *Cats: Smithsonian Answer Book*. Smithsonian Books, Washington, DC.

Sinclair, A. R. E., S. Mduma, and J. S. Brashares. 2003. Patterns of predation in a diverse predator-prey system. *Nature* 425:288–290.

Sliwa, A. 1996a. Pleasures and worries of a black-footed cat field study in South Africa. *Cat Times* 23:1–3.

Sliwa, A. 1996b. Small-spotted cat (black-footed cat). In *Smithers' Mammals of Southern Africa*, ed. P. Apps, 265–266. Southern Book Publishers, Halfway House.

Sliwa, A. 1997. Black-footed cat field research. *Cat News* 27:20–21.
Sliwa, A. 1998. Africa's smallest feline: The black-footed cat, five years of research. *Endangered Wildlife* 28:10–13.
Sliwa, A. 1999. Stalking the black-footed cat: Secrets in the night. *International Wildlife* 29 (3): 38–43.
Sliwa, A. 2000. Black-footed cat: A mighty mite. *National Geographic Magazine, Earth Almanac Section.*
Sliwa, A. 2004a. Black-footed cats: Spotted nocturnal hunters. *African Wildlife* 58 (3): 16–19.
Sliwa, A. 2004b. Home range size and social organisation of black-footed cats *(Felis nigripes). Mammalian Biology* 69.
Sliwa, A. 2006a. Atomic kitten. The secrets of Africa's black-footed cat. *BBC Wildlife Magazine* 24 (12): 36–40.
Sliwa, A. 2006b. Black-footed cat research: WAZA Project 06016. *WAZA Magazine* 8:23.
Sliwa, A. 2006c. *Felis nigripes* Burchell. In *The Mammals of Africa* 5. *Carnivora, Pholidota, Perissodactyla*, ed. J. S. Kingdon and M. Hoffmann. In *The Mammals of Africa* 1–6, ed. J. S. Kingdon, T. Butynski, and D. Happold. Academic Press, Amsterdam.
Sliwa, A. 2006d. Seasonal and sex-specific prey-composition of black-footed cats *Felis nigripes*. *Acta Theriologica* 51 (2): 195–206.
Sliwa, A., M. Herbst, and M. Mills. Forthcoming. Black-footed cats *(Felis nigripes)* and African wild cats *(Felis silvestris):* A comparison of two small felids from South African arid lands. In *Felid Biology and Conservation*, ed. D. Macdonald and A. Loveridge. Oxford University Press, Oxford.
Sousa, K. S., and A. Bagar. 2007. Feeding habits of Geoffroy's cat *(Leopardus geoffroyi)* in southern Brazil. *Mammalian Biology*: 1–6.
Spencer, N. 2007. *The Gayer-Anderson Cat*. The British Museum Press, London.
Sunquist, M., and F. Sunquist. 2002. *Wild Cats of the World*. The University of Chicago Press, Chicago.
Tabor, R. 1997. *Cats: The Rise of the Cat*. Kingfisher, New York.
Traylor-Holzer, K., D. Reed, L. Tumbelaka, N. Andayani, C. Yeong, D. Ngoprasert, and P. Duengkae. 2005. *Asiatic Golden Cat in Thailand: Population and Habitat Viability Assessment; Final Report.*
Trigo, T. C., T. R. O. Freitas, G. Kunzler, et al. 2008. Inter-species hybridization among neotropical cats of the genus *Leopardus*, and evidence for an introgressive hybrid zone between *L. geoffroyi* and *L. tigrinus* in southern Brazil. *Molecular Ecology* 17:4317–4333.
Tucker, S. A., W. R. Clark, and T. E. Gosselink. 2008. Space use and habitat selection by bobcats in the fragmented landscape of south-central Iowa. *Journal of Wildlife Management* 72:1114–1124.
Villalba, L., M. Lucherini, S. Walker, et al. 2004. *The Andean cat: A Conservation Action Plan*. Andean Cat Aliance, La Paz, Bolivia.
Walker, R. S., A. J. Novaro, P. Perovic, R. Palacios, E. Donadio, M. Lucherini, M. Pia, and M. S. López. 2007. Diets of three species of Andean carnivores in high-altitude deserts of Argentina. *Journal of Mammalogy* 88 (2): 519–525.

Wang, E. 2002. Diets of ocelots *(Leopardus pardalis)*, margays *(L. wiedii)*, and oncillas *(L. tigrinus)* in the Atlantic rainforest in southern Brazil. *Studies on Neotropical Fauna and Environment* 37:207–212.

Werdelin, L., N. Yamaguchi, W. E. Johnson, and S. J. O'Brien. 2010. *Felid phylogeny and evolution.* In *The Biology and Conservation of Wild Felids*, ed. D. M. Macdonald and A. Loveridge, 59–82. Oxford University Press.

Wright, P., and M. Wright. 2000. Wildcats of Arabia. *Arabian Wildlife*, Winter 2000/2001: 32–35.

Yufeng, Y., Drubgyal, Achu, L. Zhi, and J. Sanderson. 2007. First photographs in nature of the Chinese mountain cat. *Cat News* 47:6–7.

Ziesler, G. 1992. Souvenir d'un chat des Andes. *Animan, Nature et Civilisations* 50:68–79.

Index

Acinonyx jubatus. *See* cheetah
Africa, 11–12, 59–60, 77, 122–23, 125; savannas of, 7. *See also individual countries*
African golden cat *(Caracal aurata)*, 125–26, 129; color of, 42, 45; conservation of, 96; diet of, 79; distribution of, 11; ecology of 56, 58, 60–61; lineages of, 14, 16; size and function of, 21; young of, 69
agouti, 43–44. *See also* color
Air France, 100
albinism, 44. *See also* color
alleles, 43–44. *See also* DNA; genes
alligators, 63
Americas, 16. *See also* Central America; North America; South America; *individual countries and states*
Amory, Cleveland, 113
Amur leopard cat, 49
analogous species, 11
Andean cat *(Leopardus jacobita)*, xiii, 109, 122–23, 125–26, 129; color of, 42–43, 45–46, 48; conservation of, 96–97; diet of, 76–77, 79–82; distribution of, 11; ecology of, 55, 57, 60–61; lineages of, 14, 16; size and function of, 21, 34; young of, 69, 71
Andes, 6, 46, 55, 61, 97, 109, 122. *See also individual countries and species*
Antarctica, 1, 10
anus, 38. *See also* feces
aquatic cats, 30, 63
arboreal cats, 8, 10–11, 32, 39, 58, 76
Argentina, 11, 43, 109, 122
Arizona, 92
Asia, 11, 47, 59, 77, 99, 123, 125. *See also individual regions and countries*
Asiatic golden cat *(Catopuma temminckii)*, xiii, 109, 122, 129; color of, 42; conservation of, 96; diet of, 79; distribution of, 11; ecology of, 56, 60–61; lineages of, 14–15; size and function of, 21; young of, 69
Australia, 10–13, 61, 99, 122

badger, 56
balance, 6
Bali, 10, 12–13
Barbera, Joseph, 111
Bastet, 104–107
bat, 76
bay cat *(Catopuma badia)*, xiii, 123–26, 129; color of, 42; conservation of, 96; diet of, 79; distribution of, 11; ecology of, 56, 60–61; size and function of, 21; young of, 69
bay cat lineage, 14–15, 17, 129. *See also* Asiatic golden cat *(Catopuma temminckii);* bay cat *(Catopuma badia);* marbled cat *(Pardofelis marmorata)*
bear, 4, 62
behavior of cats, xiv, 53, 83–84
Beni Hasan, 105
biodiversity, 7, 10, 65, 89. *See also* CITES (Convention on International Trade in Endangered Species; extinction; IUCN Red List
biology, xiv, 123
birds, 63, 76, 79–80, 85, 95, 98. *See also individual bird species*
blackbuck (*antelope*), 84
black-footed cat *(Felis nigripes)*, xiii, 129; color of, 42; conservation of, 96; diet of, 79–80; distribution of, 11; ecology of, 56, 60–61; lineages of, 14, 16; size and function of, 21; young of, 67, 69
bobcat *(Lynx rufus)*, xiii, 123–24, 129; color of, 40, 42; conservation of, 96; diet of, 75, 79, 81; distribution of, 11; ecology of, 60–61; lineages of, 14, 16; size and function of, 21, 32, 35; young of, 69
Bolivia, 43, 46, 48, 71, 76, 83, 100, 122
bones, 3, 5, 83. *See also* skeletal frame
Borneo, 10, 12, 14–15, 47, 83, 88, 101, 123–24
Brazil, 10, 46–47
Breitenmoser, Urs, 56, 131
Breitenmoser-Würsten, Christine, 56, 131
British Museum, 105–6
Bubastis, 105
Buckley-Beason, Valerie, 48, 131
buffalo, 44

caiman, 63
Cairo, 105
California, 109
camouflage, *See* pattern
Canada, 122
Canada lynx *(Lynx Canadensis)*, 122–23, 129; color of, 42; conservation of, 96, 98; diet of, 78–79, 81; distribution of, 11; ecology of, 60–61; lineages of, 14,16; relationship with humans, 87; size and function of, 21, 32; young of, 69
Canidae, 51. *See also* dogs; wild dogs; wolves
captivity, 66, 68, 71–72, 82
caracal *(Caracal caracal)*, 91, 122, 126, 129; color of, 42; conservation of, 96; diet of, 75, 79–80; distribution of, 11; ecology of, 60–61; lineages of, 14, 16; relationship with

caracal *(Caracal caracal) (continued)*
humans, 84; size and function, 21, 28; young of, 69–70, 73
Caracal aurata. *See* African golden cat
Caracal caracal. *See* caracal
caracal lineage, 14, 16, 129. *See also* African golden cat *(Caracal aurata)*; caracal *(Caracal caracal)*; serval *(Leptailurus serval)*
carrion, 83
Carroll, Lewis, 114, 118–19
cat lineages, 13–17, 129. *See also specific families and species*
catnip, 87–88
Catopuma badia. *See* bay cat
Catopuma temminckii. *See* Asiatic golden cat
caudal glands, 38
Central Africa, 14
Central America, 11, 14, 31, 100, 122–23, 125
Cervidae, 75
cheek, 34, 36–37
cheetah *(Acinonyx jubatus)*, xiii, 7, 14–15, 78, 84, 103, 121
chickens, 76, 83, 85, 89–92, 102
Chile, 10–11, 41, 43, 48, 53, 56, 75, 91, 109, 122
Chiloé island, 10, 109
chimpanzee, 123
chin, 6, 34, 36–37
China, 11, 101
Chin A Foeng, Steven, 77
chinchillas, 76–77
Chinese mountain cat *(Felis bieti)*, 35, 41, 56, 101
chromosome, 16, 37, 59
Christianity, 106–7, 109
Churchill, Sir Winston, 113
CITES (Convention on International Trade in Endangered Species), 101–2
classification, xiv, 2, 13, 95
claw, 27–28, 81. *See also* dew-claw
Clemens, Samuel Langhorne. *See* Twain, Mark; *individual works*
climbing, 27, 31–32, 91–92
clouded leopard *(Neofelis nebulosa)*, 25, 126, 130; color of, 42, 48; conservation of, 96; diet of, 79–80, 82; distribution of, 11; ecology of, 60–61; lineages of, 14–15; size and function of, 31–33, 39; teeth of, 5; young of, 69
coat. *See* fur
coexist, 58
color, 40–42, 45–47
communication, 38, 54
Cossios, Daniel, 48
crocodiles, 63

Darwin, Charles, 12, 123
Davis, Jim, 110
Dayak, 101
deer: forest, 82; roe, 75
De Montaigne, Michel, 114
Deng, Xiaoping, 114
den, 69–71
dentition, 3–4. *See also* teeth
dermis, 40. *See also* skin
desert, 10, 60
development, xiv
dew-claw, 5, 28
diet, 75, 79. *See also* prey
disease, 64; bacteria, 92; black plague, 109; viruses, 92
distribution, geographic, 11, 59
diurnal cats, 8
DNA, 13, 37, 48, 127. *See also* genes
Dodgson, Charles Lutwidge. *See* Carroll, Lewis
dog, 4, 6, 8; guard, 91. *See also* wild dogs
domestic cat *(Felis catus)*, xiii, xiv, 1, 7, 9–10, 13, 109, 122, 129; behavior of, 53; color of, 42; diet of, 76, 79; distribution of, 11; ecology of, 61; lineages of, 14, 16; size and function of, 21, 23, 36; young of, 67, 69
Dr. Seuss, 109

eagle, 63
ear, 81; auditory bullae, 4
ecosystem, 65
Ecuador, 46, 100
Egypt, 104–7, 109. *See also* Bastet; mummified cats
Eizirik, Eduardo, 43
Elliot, Daniel Giraud, 124
Eliot, T. S., 117
endemic, 15. *See also* distribution of cats
enemies of cats, 63
environmental, 32
epidermis, 40
ESA (Endangered Species Act), 86–87
estrogen, 68
Eurasian lynx *(Lynx lynx)*, 8–9, 122–24, 129; behavior of, 52; color of, 42; conservation of, 96, 98; diet of, 75, 79, 81; distribution of, 11; ecology of, 57, 60–61; lineages of, 14, 16; size and function, 21–23, 25, 32; young of, 69, 74
Europe, 11, 59–60, 122. *See also individual countries*
evolution, xiv, 3, 7, 12, 27, 45, 47, 49, 52
extinction, 95, 98–99
eye, 26, 44. *See also* vision

Fauna Communications Institute of Hillsborough, 64
feces, 48, 51, 55, 127
Felidae, xiii, 1–2, 7, 9, 13–15, 17, 22, 37, 39, 99, 125
Felis lineage, 14, 16–17, 129. *See also* black-footed cat *(Felis nigripes)*; domestic cat *(Felis catus)*; jungle cat *(Felis chaus)*; sand cat *(Felis margarita)*; wildcat *(Felis silvestris)*

Felis catus. *See* domestic cat
Felis chaus. *See* jungle cat
Felis margarita. *See* sand cat
Felis nigripes. *See* black-footed cat
Felis silvestris. *See* wildcat
Felix the Cat, 109–10
feral cats, xiii, 69, 85, 79, 99
fish, 30, 76, 79, 87, 125
fishing cat *(Prionailurus viverrinus)*, 98, 125–26, 129; color of, 42; conservation of, 96; diet of, 76–77, 79; distribution of, 11; ecology of, 60–61; lineages of, 13–14, 16; size and function of, 21, 30; tail of, 6; young of, 69
flat-headed cat *(Prionailurus planiceps)*, xiii, 98, 123, 125–26, 129; color of, 42; conservation of, 96; diet of, 76, 79; distribution of, 10–11; ecology of, 56, 60–61; lineages of, 13–14, 16; size and function of, 21, 30; tail of, 6; young of, 69
flehmen, 37
Fleischer, Lenore, 120
forests: boreal, 8; temperate, 11; tropical, 10–11, 61
fossils, 17–19, 39, 47
foxes, 8
France, 19
frogs, 6, 11, 30, 76, 79, 85
fungi, 92
fur, 40–41, 43–44, 51

Galapágos, 10
Garfield (Davis), 110–11
Gautier, Theophile, 113
Gayer-Anderson, Major Robert Grenville, 105
Gayer-Anderson Museum, 105–6
geese, 76, 85, 89, 91, 102
genes, 43–44, 48, 86. *See also* DNA
genitals, 68. *See also* reproduction
Geoffroy's cat *(Leopardus geoffroyi)*, 129; behavior of, 53; color of, 40, 42, 48; conservation of, 96; diet of, 79; distribution of, 11; ecology of, 59–61; interaction with humans, 85; lineages of, 14, 16; size and function of, 21; young of, 69
geographic range, xiii, 49, 99, 125
geographic variation, 47–49
global warming, 97–98
Gondwanaland, 12
Grand Canyon National Park, 92
grassland, 10, 58, 61
Grassman, Lon, 109, 131
Great Basin desert, 60
Great Family of Cats, xiii
Greece, cats in ancient, 104
Gregory the Great (pope), 107
guigna *(Leopardus guigna)*, 109, 123, 126, 129; color of, 40–42, 48; conservation of, 96; diet of, 75–76, 79, 83; distribution of, 11; ecology of, 53–54, 56, 60–61; interaction with humans, 85, 89, 91–92; lineages of, 14, 16; size and function of, 20, 23, 27; young of, 69
gymnastics, 31

habitat, 7–8, 10–11, 45, 58, 61, 87, 125
hair. *See* fur
Hanna, William, 111
Hawaii, 10
heat, 66, 71
heart, 22–23
henhouse, 83, 91
herbivores, 64
Herodotus, 104–5
hibernate, 62–63
Hindus, 106
Hobbs, Thomas, 73
home range, 50–51, 58
humans, xv, 3, 6, 10, 20, 52, 54, 56, 80–81, 83, 121; interaction with cats, 85–87, 89–93, 98–99, 101–3
hummingbird, 62
Huxley, Sir Julian, 113
hyenas, 19
hypercarnivores, 1, 90

Iban, 101
Iberian lynx *(Lynx pardinus)*, 122–23, 126, 129; color of, 42; conservation of, 95–96; diet of, 78–79; distribution of, 11; ecology of, 60–61; lineages of, 14, 16; size and function of, 21, 32; young of, 69, 72
Ice Age, 44
India, 12, 21, 80, 84, 88
Indian antelope, 84
infanticide among cats, 63–64
intelligence, 51–52
Iriomote island, 48–49
Isla Grande de Chiloé, 20. *See also* island gigantism
IUCN (International Union for Conservation of Nature), 78, 94–97, 101

jackals, 8
jackrabbits, 88
Jacobson's organ, 37
jaguar *(Panthera onca)*, 7, 9, 14–15, 18, 58, 121
jaguarundi *(Puma yagouaroundi)*, 126, 130; color of, 42; conservation of, 96; diet of 77, 79; distribution of, 11; ecology of, 58, 60–61; lineages of, 14–15; size and function of, 21, 34; young of, 69
Japan, 49, 110
Java, 10, 12, 47, 125
jaw, 2–5, 36, 76
jumping, 28–29, 75
jungle cat *(Felis chaus)*, 129; color of, 42; conservation of, 96; diet of, 79–80; distribution of, 11; ecology of, 60–61; lineages of, 14, 16; size and function of, 21, 28; young of, 69

Kipling, Rudyard, 120
kittens, 46, 52, 63–64, 70–73, 84–85, 88. *See also* den; litter

La Paz, 100
latrine, 55
leopard *(Panthera pardus)*, xiii, 7, 12, 14, 103, 121
leopard cat *(Prionailurus bengalensis)*, 122, 129; color of, 42, 45, 47–48; conservation of, 96; diet of, 77, 79–80, 83; distribution of, 11; ecology of, 60–61; lineages of, 12 14, 16–18; size and function of, 21; young of, 69
leopard cat lineage, 14, 16–18, 129. *See also* fishing cat *(Prionailurus viverrinus)*; flat-headed cat *(Prionailurus planiceps)*; leopard cat *(Prionailurus bengalensis)*; Pallas' cat *(Otocolobus manul)*; rusty-spotted cat *(Prionailurus rubiginosus)*
Leopardus, 17
Leopardus colocolo. *See* pampas cat
Leopardus geoffroyi. *See* Geoffroy's cat
Leopardus guigna. *See* guigna
Leopardus jacobita. *See* Andean cat
Leopardus pardalis. *See* ocelot
Leopardus tigrinus. *See* tigrina
Leopardus wiedii. *See* margay
Leptailurus serval. *See* serval
Leviathan, 73
Leyhausen, Paul, 76, 134
ligaments, 27
Lincoln, Abraham, 114
lineages, cats. *See* cat lineages; *individual cat lineages*
Linne, Carl von (Linnaeus), 9–10
lion *(Panthera leo)*, xiii, 7, 9, 14–15, 48, 55, 103
litter, 71
livestock, 79, 91, 103
lizard, 41, 76, 79, 85
Lombok, 10, 12–13
Looney Tunes (Warner Brothers), 112
Los Gatos, 109
Lucifer, 109
lucky cat, 109
Lynx canadensis. *See* Canada lynx
Lynx lynx. *See* Eurasian lynx
Lynx pardinus. *See* Iberian lynx
Lynx rufus. *See* bobcat
lynx lineage, 14, 16–17, 123, 129. *See also* bobcat *(Lynx rufus)*; Canada lynx *(Lynx Canadensis)*; Eurasian lynx *(Lynx lynx)*; Iberian lynx *(Lynx pardinas)*

Madagascar, 10–12
Mafdet, 104
Maine, 60
Malay language, 25, 109
Malay Peninsula, 10
McGill University, Montreal, Canada, 118
mammals, 1, 12–13, 45–46, 56, 62, 64, 79, 69
maneki neko, 109–10
map of small cat species, 11
marbled cat *(Pardofelis marmorata)*, 1, 125–26, 129; color of, 42; conservation of, 96; diet of, 76, 79; distribution of, 11; ecology of, 56, 60–61; lineages of, 14–15; size and function of, 21,31–32; tail of, 6; young of, 69
margay *(Leopardus wiedii)*, 126, 130; color of, 42, 46; conservation of, 96; diet of, 76, 79; distribution of, 11; ecology of, 58–61; interaction with humans, 85; lineages of, 14, 16; size and function of, 21, 31–32; young of, 69
marmots, 56
marking, 42. *See also* color
Marx, Groucho, 113
Maryland, 15
Mayr, Ernst, 47
Mediterranean, 105
melanin, 40–41, 43
metabolism, 62
Merrie Melodies (Warner Brothers), 112
Metro-Goldwyn-Mayer, 111
Mexico, 60, 123
migration, 56, 58
Middle East, 8, 10, 99, 103, 125
Mitchard, Jacquelyn, 113
Mivart, St. George, 107, 134
molecular analysis of cat lineages, 1, 15, 124. *See also* DNA; genes
Mongolia, 62, 101, 122–23
morphology, 1, 13, 47–48
mouth and throat: hyoid, 53; larynx, 53; papillae, 37–39, saliva, 53. *See also* dentition; jaws; teeth
mummified cats, 105, 107
muscles, 8, 64, 73, 81
Muslims, 106
muzzle, 6
mythology, 103–4

Napolitano, Constanza, 23, 66, 76, 135
National Geographic, 107
National Institute of Health in Fredericksburg, 15
Native Americans, 44, 100
Neofelis, 14–15, 130
Neofelis diardi. *See* Sunda clouded leopard
Neofelis nebulosa. *See* clouded leopard
Nepal, 121
NGO (non-governmental organizations), 95
New Guinea, 10, 12
New Mexico, 88
New Zealand, 12
nine lives of cats, 86, 107
nocturnal cats, 8, 58, 83, 85
North Africa, 8, 10, 43, 125
North America, 14–15, 40, 59, 122–23

North Carolina, 64
Northern Hemisphere, 10–11

O'Brien, Stephen, 15, 43, 135
obligate, 1, 90
ocelot *(Leopardus pardalis)*, xiv, 122–23, 126, 130; color of, 42, 45; conservation of, 96, 101; diet of, 77, 79; distribution of, 11; ecology of, 58–61; lineages of, 14, 16–17; size and function of, 20–21, 36; young of, 67, 69
ocelot lineage, 14, 16–17, 20, 45, 130. *See also* Andean cat *(Leopardus jacobita);* Geoffroy's cat *(Leopardus geoffroyi); guigna (Leopardus guigna);* margay *(Leopardus wiedii);* ocelot *(Leopardus pardalis);* pampas cat *(Leopardus colocolo);* tigrina *(Leopardus tigrinus)*
Oligocene, 12, 18
omnivores, 64
Otocolobus manul. See Pallas' cat
otter, 4
owl, 63

Palawan, 10
Pallas' cat, 122–23, 129; color of, 42, 45–46; conservation of, 96, 98, 101; diet of, 78–80; distribution of, 11; ecology of, 56, 60–61; lineages of, 14, 16; size and function of, 21; young of, 69, 73
palm oil, 30, 61, 77, 83
pampas cat *(Leopardus colocolo)*, 122, 130; color of, 40, 42, 46–47; conservation of, 96; diet of, 79; distribution of, 11; ecology of, 60–61; lineages of, 14, 16; size and function of, 21; young of, 69
Panthera, 15
Panthera leo. See lion
Panthera onca. See jaguar
Panthera pardus. See leopard
Panthera tigris. See tiger
Panthera uncia. See snow leopard
Panthera lineage, 14–15, 130. *See also* clouded leopard *(Neofelis nebulosa);* jaguar *(Panthera onca);* leopard *(Panthera pardus);* lion *(Panthera leo);* snow leopard *(Panthera uncia);* Sunda clouded leopard *(Neofelis diardi);* tiger *(Panthera tigris)*
papillae, 37–39
parasites, 92
Pardofelis marmorata. See marbled cat
Paris, 100
patterns, 42, 45–47. *See also* camouflage
paws, 5–7, 28, 62
Peru, 43, 46, 48, 100, 109
pest, cats as, 90–91
Phaeomelanin, 41
Philippines, 10, 47, 49
pikas, 78–79
play, 52–53
Pluto, 119
Poe, Edgar Allan, 119
polar bear, 44
population, 10, 49, 52, 78, 94–95, 102, 125–26
Portugal, 78
poultry, 79, 85, 99
predators, cats as, xiii, 1, 7–8, 10, 39, 54, 56, 63–64, 75–77, 84, 87, 89
prey, 2, 5–6, 8, 27, 65, 75, 78–81, 82–83, 90, 97, 124. *See also* diet; *individual prey*
primates, 12, 79, 82, 95; chimpanzee, 123; gorilla, 46, 123; monkey, 79
Prionailurus bengalensis. See leopard cat
Prionailurus planiceps. See flat-headed cat
Prionailurus rubiginosus. See rusty-spotted cat
Prionailurus viverrinus. See fishing cat
Proailurus, 17–19
Ptolemy xii, 104
pudú, 76
puma *(Puma concolor)*, 7, 14–15, 46, 58, 92, 121
puma lineage, 14–15, 17, 130. *See also* cheetah *(Acinonyx jubatus);* jaguarundi *(Puma yagouaroundi);* puma *(Puma concolor)*
Puma yagouaroundi. See jaguarundi
purring, 7, 64

rabies, 92
radio collar, 83, 89
rats, 77, 79, 83
Red List (IUCN), 95–96, 101
region, 103, 105
religion, 103–6
Renaldi, Alcides, 76
reproduction, xiv; gestation, 67–69; mating, 50, 66–68; ovulation, 66; pregnancy, 67
righting reflex, 108
rodents, 7, 10, 77, 79, 85, 89, 91–92, 103
Rogers, Will, 113
Russia, 123
rusty-spotted cat *(Prionailurus rubiginosus)*, 129; color of, 42; conservation of, 96; diet of, 79–80; distribution of, 11; ecology of, 60–61; lineages of, 14, 16; size and function of, 21–22, 24; young of, 69

sand cat *(Felis margarita)*, 126, 129; color of, 42; conservation of, 96; diet of, 79, 83; distribution of, 10–11; ecology of, 60–61; lineages of, 14, 16; size and function of, 21; young of, 69
Sarawak, 110–11, 124
savannas, 11, 58, 77–78, 80
Saxony, 107
scavenging, 83
scent glands, 38
senses, 6–7, 36–37, 87
serval *(Leptailurus serval)*, 122,129; color of, 42; conservation of, 96; diet of, 76–80; distribution of, 11; ecology of, 58, 60–61; lineages of,

14, 16; size and function of, 7, 21, 28–29, 31; young of, 69
Shirus, George, 88
Siberian tiger *(Panthera tigris altaica)*, 8, 22
Siculus, Diodorus, 104
size of small cats, 20–22
skeleton, 3
skeletal frame, 73
skin, 93, 100–101, 124; epidermis, 40
skull, 1–5, 8, 13, 39, 81, 124
sleep, 55–57
Sliwa, Alex, 80, 123, 136–137
snakes, 77, 79, 103
snow leopard (*Panthera uncia)*, 7, 14–15, 121
Sonora desert, 60
South Africa, 80
South America, 6, 11, 31, 47, 55, 59, 97, 100, 122–23
Southeast Asia, xiii, 6, 9–12, 15, 22, 31, 45, 48, 100, 122
Southwest Asia, 12
Spain, 78, 122
species, 1, 6–7, 9–10, 98–99, 121–23, 125; behavior of, 54; color of, 46, 49; conservation of, 94–95, 101–2; ecology of, 58–60, 64–65; interaction with humans, 86–87
sperm, 68
spinal cord, 2, 3
Squier, M. L, 117
Sri Lanka, 21–22, 80
SSC (Species Survival Commission), 94–95
St-Gérand-le-Puy, 19
Sulawesi, 10, 12
Sumatra, 10, 12, 47, 122
Sunda clouded leopard *(Neofelis diardi)*, 126, 130; color of, 42, 48; conservation of, 69; diet of, 79; distribution of, 11; ecology of, 60–61; lineages of, 14–15; size and function of, 21, 31, 39; young of, 69
Suriname, 58, 77
swamps, 10
swimming, 27, 76
Switzerland, 56, 94, 107
Sylvester the Cat, 112

tails, 6, 33–34
tapetum luvidum, 26
taxonomy, 13
Taylor-Still, Andrew, 114
teeth, 1–5, 13, 39, 81–82; diastema, 2. *See also* dentition
temperature, 62
terrestrial cats, 7, 39, 54, 58, 63, 67
territories, 50, 66–67, 74
testosterone, 68
Thailand, 109
Tibetan plateau, 10
tiger *(Panthera tigris)*, xiii, 7, 9, 12, 14–15, 18, 44, 48
tigrina *(Leopardus tigrinus)*, 125–26, 130; color of, 40, 42; conservation of, 96; diet of, 79; distribution of, 11; ecology of, 58–61; interaction with humans, 85; lineages of, 14, 16; size and function of, 21–22; young of, 69
titi, 100, 110
toes, 5–6
Tom and Jerry (Metro-Goldwyn-Mayer), 111
Tom Sawyer (Twain), 114
tongue, 37–38
trail camera, 88–89
traps, 100
Trinidad, 10

United States, 87, 99, 111–12, 122
urine, 38, 51

Van Vechten, Carl, 113
vectors, 92
vegetables, 103
vegetation, 55, 64–65
Veterinary Association, 64
Verne, Jules, 114
vibrissae, 6–7, 32, 34–36, 37, 87. *See also* whiskers
vicunas *(Vicugna vicugna)*, 55
Villalba, Lilian, 76
viscachas, 76, 80–82, 97–99
vision, 24, 26. *See also* eyes; senses
vocalization, 66. *See also* purring
vomeronasal, 37

Wales, 107
Walker, Susan, 76
Wallace, Alfred Russel, 12–13, 123
Wallace Line, 10, 13
Warner Brothers, 112
water, xiii, 8, 27, 60, 76, 83, 125
Werdelin, Lars, 18, 138
Western Hemisphere, 10–11
whiskers, 6, 32, 34–37. *See also* vibrissae
Whitehead, Alfred North, 113
wildcat *(Felis silvestris)*, 103, 122–23, 129; color of, 42; conservation of, 96; diet of, 79; distribution of, 11; ecology of, 56, 60–61; interaction with humans, 85; lineages of, 16; size and function of, 21; young of, 69
wolves, 8
wild dogs, 2, 4, 6, 8. *See also* dogs
Wordsworth, William, 117
wrist, 6, 34